Numerical Simulation
of 3-D Incompressible
Unsteady Viscous Laminar Flows

Edited by Michel Deville, Thien-Hiep Lê,
and Yves Morchoisne

Notes on Numerical Fluid Mechanics (NNFM) Volume 36

Series Editors: Ernst Heinrich Hirschel, München
Kozo Fujii, Tokyo
Bram van Leer, Ann Arbor
Keith William Morton, Oxford
Maurizio Pandolfi, Torino
Arthur Rizzi, Stockholm
Bernard Roux, Marseille

(Adresses of the Editors: see last page)

Numerical Simulation of 3-D Incompressible Unsteady Viscous Laminar Flows

A GAMM-Workshop

Edited by
Michel Deville, Thien-Hiep Lê,
and Yves Morchoisne

Die Deutsche Bibliothek – CIP-Einheitsaufnahme

Numerical simulation of 3 D incompressible unsteady viscous laminar flows: a GAMM workshop / ed. by Michel Deville...
– Braunschweig; Wiesbaden: Vieweg, 1992
  (Notes on numerical fluid mechanics; Vol. 36)
  ISBN 978-3-663-00071-6   ISBN 978-3-663-00221-5 (eBook)
  DOI 10.1007/978-3-663-00221-5
NE: Deville, Michel [Hrsg.]; Gesellschaft für Angewandte Mathematik
  und Mechanik; GT

Produced by W. Langelüddecke, Braunschweig
Printed on acid-free paper

ISSN 0179-9614
ISBN 978-3-663-00071-6

FOREWORD

The GAMM-Committee for Numerical Methods in Fluid Mechanics (GAMM-Fachausschuss für Numerische Methoden in der Strömungsmechanik) has sponsored the organization of a GAMM Workshop dedicated to the numerical simulation of three-dimensional incompressible unsteady viscous laminar flows to test Navier-Stokes solvers. The Workshop was held in Paris from June 12th to June 14th, 1991 at the Ecole Nationale Supérieure des Arts et Métiers.

Two test problems were set up. The first one is the flow in a driven-lid parallelepipedic cavity at Re = 3200 . The second problem is a flow around a prolate spheroid at incidence. These problems are challenging as fully transient solutions are expected to show up. The difficulties for meaningful calculations come from both space and temporal discretizations which have to be sufficiently accurate to resolve detailed structures like Taylor-Görtler-like vortices and the appropriate time development.

Several research teams from academia and industry tackled the tests using different formulations (velocity-pressure, vorticity-velocity), different numerical methods (finite differences, finite volumes, finite elements), various solution algorithms (splitting, coupled, ...), various solvers (direct, iterative, semi-iterative) with preconditioners or other numerical speed-up procedures. The results show some scatter and achieve different levels of efficiency.

The Workshop was attended by about 25 scientists and drove much interaction between the participants. The contributions in these proceedings are presented in alphabetical order according to the first author, first for the cavity problem and then for the prolate spheroid problem.

No definite conclusions about benchmark solutions can be drawn. Indeed, because of the intricated nature of the time and space behaviours of the solutions, more work remains to be done in this area to achieve successful results.

The organizers of the Workshop would like to acknowledge the support from : Gesellschaft für Angewandte Mathematik and Mechanik (GAMM, Germany), Direction des Recherches Etudes et Techniques (DRET, Paris), Centre National de la Recherche Scientifique (CNRS, Paris), CRAY Research France, Office National d'Etudes et de Recherches Aérospatiales (ONERA, Châtillon, France) and Ecole Nationale Supérieure des Arts et Métiers.

May 19th, 1992.   MICHEL DEVILLE,

THIEN HIEP LE,

YVES MORCHOISNE.

CONTENTS

Numerical Simulation of 3-D Incompressible Unsteady Viscous Laminar Flows :

The Test Problems

M.Deville (*), T.H.Lê (**), Y.Morchoisne (**)

The purpose of the workshop was to compare numerical methods used for direct solution of the time dependent incompressible Navier-Stokes equations (DNS) and to evaluate their performances. Hereafter, the reader will find the description of each test case and the required results from the participants.

The Navier-Stokes equations are written in dimensionless form with obvious definition of the Reynolds number in each particular problem.

Symbol List

ρ (=1)	density
ν	kinematic viscosity
U	velocity field
P	static pressure field
Re (=1/ν)	Reynolds number
ω	vorticity field
Ψ	vector stream function
u,v,w,	velocity components
$\omega_x, \omega_y, \omega_z$	vorticity components

----

*   U.C.L., Louvain-la-Neuve, Belgium.
**   O.N.E.R.A., Châtillon, France.

TEST PROBLEMS

- Case 1

The first case is a driven cavity at Re=3200. The geometry is a parallelepiped such that $(x,y,z) \in [-1/2,+1/2]^2 \times [-3/2,+3/2]$. The initial condition is that of a fluid at rest :

$$U = P = \omega = \ldots = 0 .$$

The boundary conditions are no slip fixed walls everywhere, except that the top wall moves in its own plane at unit velocity in the positive x direction:

$$U = (0,0,0) \quad \text{for} \quad x = \pm 1/2 \quad \text{or} \quad y = -1/2 \quad \text{or} \quad z = \pm 3/2$$

$$U = (1,0,0) \quad \text{for} \quad y = +1/2 .$$

- Case 2

The second case is the flow over a prolate spheroid at Re=500. The geometry of the body is defined by the relationship:

$$x^2 + (y^2 + z^2)/ a^2 = 1/4 \quad \text{with} \quad a = 1/6 .$$

The initial condition is that of a fluid at rest.
The boundary conditions are:
- upstream uniform velocity : $U_\infty = (\cos \alpha , \sin \alpha)$ with $\alpha = 30°$,
- downstream : conditions chosen by the user.

QUESTIONS ON METHOD AND COST

Method

.formulation
.discretization in time and space
.stability
.boundary conditions (natural, artificial, outflow)
.time step
.space grid .

Computing Cost For Each Case

.computer used
.total cost
.number of degrees of freedom
.number of time steps .

RESULTS : CURVES

- CASE 1

As P is defined up to an arbitrary additive constant, we impose:

P = 0   at point (x = 0, y = -1/2, z = 0).

- For t = 50 , 100 and 200   give

. u(y)   for   z = 0   and   x = 0

. v(x)   for   z = 0   and   y = 0

. P(x), ω_z(x)   for   z = 0   and   y = 1/2

. P(z), ω_x(z)   for   x = 0   and   y = 0 .

- For (x = 0, y = 1/2, z = 0)   give

. P(t), ω_z(t).

- CASE 2

As P is defined up to an arbitrary additive constant, we impose:

Lim P (x = -L cos α, y = -L sin α, z = 0, t) = 0   for   L → ∞.

- For t = 1 , 2 , 3 , 4 and 5   give

. P(x), ω_z(x)   for   z = 0   and $x^2 + y^2 / a^2 = 1/4$.

- For (x = -1/2, y = 0, z = 0)   give

. P(t), ω_z(t).

RESULTS : Iso-value Contours

Iso-P, iso-ω normal component, with, for each variable and each figure, 11 levels:

$$\Phi = \Phi_0 + i \ (\ \Phi_1 - \Phi_0 \ ) \ / \ 10$$

with $\Phi = P$ or ω normal component.

Φ_0 and Φ_1 have to be choosen by the authors and given for each figure.

- Case 1

- for t = 50 , 100 and 200

. plane x = 4/15   : ω_x and P

. plane x = 0   : ω_x and P

. plane x = -7/15 : ω_x and P

. plane y = 1/2   : ω_z and P

. plane z = 0   : ω_z and P.

- Case 2

- for t = 1 , 3 and 5

. plane x = 0   : ω_x, P for $-.25 \leq y \leq .25$ and $-.25 \leq z \leq .25$

. plane x = .375   : ω_x, P for $-.25 \leq y \leq .25$ and $-.25 \leq z \leq .25$

. plane x = -.375 : ω_x, P for $-.25 \leq y \leq .25$ and $-.25 \leq z \leq .25$

. plane y = 0   : ω_y, P for $-1. \leq x \leq 1.$ and $-1. \leq z \leq 1.$

. plane z = 0   : ω_z, P for $-1. \leq x \leq 1.$ and $-1. \leq y \leq 1.$

REFERENCES

EXPERIMENTS FOR THE 3-D DRIVEN CAVITY

[01] J.R.Koseff, R.L.Street, P.M.Gresho, C.D.Upson,
J.A.C.Humphrey, W.M.To,
"A 3-D driven cavity flow: experiment and simulation",
Proc. 3rd Int.Conf.on Num.Meth. for Lam. and Turb. flow,
Univ. Washington, pp 564-581, Aug. 8-11, 1983.

[02] J.R.Koseff, R.L.Street,
"Visualization studies of a shear driven 3-D recirculating
flow", J. Fluids Engin., 106, 21-29, 1984.

[03] H.S.Rhee, J.R.Koseff, R.L.Street,
"Flow visualization of a recirculating flow by rheoscopic
liquid and liquid crystal techniques",
Exp. Fluids, 2, 57-64, 1984.

NUMERICAL SIMULATIONS FOR THE 3-D DRIVEN CAVITY

[04] J.Kim, P.Moin,
"Application of a fractional-step method to incompressible
Navier-Stokes equations", J.Comput.Phys., 59, 308-323, 1985.

[05] C.J.Freitas, A.N.Findikakis, R.L.Street, J.R.Koseff,
"Numerical simulation of 3-D flow in a cavity",
Int. J. Num. Meth. Fl., 5, 561-575, 1985.

[06] C.Y.Perng, R.L.Street,
"3-D unsteady flow simulations : alternative strategies for
a volume-averaged calculation",
Int. J. Num. Meth. Fl., 9, 341-362, 1989.

[07] M.Hafez, M.Soliman,
"A velocity decomposition method for viscous incompressible
flow calculations", AIAA 89-1966-CP, 9th Comp. Fluid
Dynamics Conference, Buffalo, N.Y., June 13-15, 1989.

REFERENCES

EXPERIMENTS FOR THE PROLATE SPHEROID

[08] H.U.Meier, H.P.Kreplin,
"Experimental investigation of the boundary layer
transition and separation on a body of revolution",
Z. Flugwiss. Weltraumforsch., 4, Heft 2, 65-71, 1980.

[09] H.Vollmers, H.P.Kreplin, H.U.Meier,
"Separation and vortical-type flow around a prolate
spheroid: evaluation of relevant parameters",
AGARD CP 342, pp 14-1, 14-4, 1983.

NUMERICAL SIMULATIONS FOR THE PROLATE SPHEROID

[10] T.C.Tai,
"Determination of 3-D flow separation by streamline method",
AIAA J., 19, 1264-1271, 1981.

[11] V.C.Patel, J.H.Baek,
"Boundary layers and separation on a spheroid at incidence",
AIAA J., 23, 55-63, 1985.

[12] V.N.Vatsa, J.L.Thomas, B.W.Wedan,
"Navier-Stokes computations of prolate spheroid at angle of
attack", AIAA 87-2627 CP, Monterey, Ca, USA, Aug. 87.

[13] T.C.Wong, O.A.Kandil, C.H.Liu,
"Navier Stokes computations of separated vortical flows past
prolate spheroid at incidence", AIAA 89-0553 27th Aerospace
Sciences Meeting, Reno, Nevada, USA, Jan. 9-12, 1989.

THE CHALLENGES OF THE NUMERICAL INTEGRATION OF THE TRANSIENT THREE-DIMENSIONAL NAVIER-STOKES EQUATIONS

M.O.DEVILLE

Université Catholique de Louvain, Unité de Mécanique Appliquée
Louvain-la-Neuve, Belgium
T.H. LE, Y.MORCHOISNE
ONERA, Châtillon, France

The purpose of the worshop is the evaluation of numerical methods for the integration of the unsteady three dimensional incompressible Navier-Stokes equations at moderate Reynolds number. This is a cornerstone for direct numerical simulation intended to turbulent flows. In this introduction, we like to address and summarize a few topics devoted to this kind of numerical problems.

1 The continuous formulation

Although the classical Navier-Stokes equations are known since the previous century, the intrusion of numericists in the field of classical fluid mechanics made an impact on which equations should be solved.

The first approach resorts to velocity-pressure formulation :

$$\rho \left(\frac{\partial \underline{v}}{\partial t} + \underline{v} \cdot \underline{\nabla} \underline{v} \right) = - \nabla p + \mu \Delta \underline{v} , \tag{1}$$

$$\nabla \cdot \underline{v} = 0, \tag{2}$$

where ρ is the density , $\underline{v}$ the velocity field, p the pressure and μ the dynamic viscosity. Eq. (1) is the momentum equation while Eq. (2) imposes the continuity constraint or the incompressibility condition.

Another formulation is based on velocity-vorticity variables :

$$\rho \left(\frac{\partial \omega}{\partial t} + \underline{v} \cdot \underline{\nabla} \omega \right) = \underline{\omega} \cdot \underline{\nabla} \underline{v} + \mu \Delta \underline{\omega}, \tag{3}$$

7

$$\Delta v = - \text{curl } \omega, \tag{4}$$

$$\omega = \text{curl } v, \tag{5}$$

where ω defined by Eq. (5) denotes the vorticity. Eqs. (3) and (4) are favored by numericists who want to get rid of the pressure. When the vorticity is used as dependent variable, the continuity equation (2) is automatically satisfied in the continuous formulation. This should be checked in the discrete equations.

Finally, a third formulation is based on vorticity-potential vector approach. Here, the continuity constraint is automatically satisfied and the pressure is no longer present in the equations. We have :

$$\rho \left(\frac{\partial \omega}{\partial t} + \underline{v} \cdot \nabla \omega \right) = \underline{\omega} \cdot \nabla \underline{v} + \mu \Delta \omega, \tag{6}$$

$$\Delta \psi = - \omega, \tag{7}$$

$$\underline{v} = \underline{\text{curl }} \psi , \tag{8}$$

where the potential vector ψ is related to the velocity field by the definition (8). From (8), it is easily seen that the stream-function formulation satisfies condition (2).

The velocity-pressure formulation is often rewritten in such a way that the pressure is obtained from a Poisson equation. Applying the divergence operator to Eq.(1) and taking Eq.(2) into account, we obtain :

$$\rho \left(\frac{\partial \underline{v}}{\partial t} + \underline{v} \cdot \nabla \underline{v} \right) = - \nabla p + \mu \Delta \underline{v} . \tag{9}$$

$$\Delta p = - \rho \, \underline{\nabla} \cdot (\underline{v} \cdot \underline{\nabla} \, \underline{v}). \tag{10}$$

The solution of (10) raises the question of the pressure boundary conditions.

The two other approaches (velocity-vorticity and vorticity-potential vector) avoid the pressure problem. However, in order to integrate Eqs (3) or (6), we need to impose vorticity boundary values which are derived quantities through Eqs.(5) or (7).

2 The discrete formulation.

Once the continuous equations have been chosen, the numericist faces the choice of the solution functional space.

If we use the strong formulation for Eqs.(9) and (10), we look for solutions of $\underline{v}$ and p in H^2, the Sobolev space containing functions and derivatives up to order two which are square integrable. However, a weak formulation allows for the search of $\underline{v}$ and p in H^1 and L^2, respectively. This is an advantage from a computational point of view [5].

2.1 Space discretization

Finite differences (FD), finite volumes (FV) [12], finite elements (FE) [5,8] and spectral approximations [2] are candidates for the space discretization. They tackle the problem in strong formulation (finite differences, spectral schemes) or in a variational form (finite elements, spectral elements). The finite volumes constitute a special form of integral approach. Most of the approximation schemes (FD, FV, FE) are algebraically convergent (h-type) and of second-order accuracy. Some of them are exponentially convergent as spectral schemes or p-type elements in absence of singularities.

Among the various discretizations, the construction of the grid is a major ingredient of any numerical technique. The grids may be staggered or non-staggered. In the latter case, experienced numericists know how to cope with the presence of spurious pressure modes, a subject of abundant research.

2.2 Time discretization

Time marching schemes may be explicit, implicit or semi-implicit [4]. Most of them are second-order accurate. The pros and cons for each choice are numerous and subject to many considerations. From the point of view of the practioner, explicit schemes are easy to implement, vectorize and parallelize well. Their major drawback is their conditional stability which is driven essentially by the Courant condition at high Reynolds numbers. Implicit schemes are unconditionally stable but lead to the solution of large linear systems. They are difficult to program and are the algorithmic bottleneck of such a treatment. They require very efficient solvers based on good preconditioners

[6]. Therefore, semi-implicit schemes (predictor-corrector, semi-implicit Runge-Kutta, etc) offer a viable and interesting choice. We should remind that solving the unsteady Navier-Stokes equations consists in solving a stiff problem, especially for advection dominated flows. Therefore, the numerical time scheme has to possess good stability properties [15].

3 The pressure computation

Let us restrict ourselves in this section to the velocity-pressure formulation. For two-dimensional flows, it is possible to solve Eqs.(1) and (2) in a coupled implicit way. In this case, the pressure is the physical variable which is directly controlled by the continuity constraint. The extension of this coupled approach to three-dimensional problems yields a very large algebraic system which is almost untractable on present day computers. A split form of the Navier-Stokes equations is often used, where a pressure Poisson equation (10) must be solved. One of the key questions is the nature of the boundary condition to apply in order to compute the pressure accurately in space *and* time [7,11].

The presence of pressure spurious modes was discovered among the various discretizations (FV, FD, spectral, FE) and several solutions to cure the problem were proposed [9]. The Galerkin statement of the finite element method produces the Brezzi-Babuska condition which is a compatibility condition for the discrete velocity and pressure spaces. Typically, the polynomial degree for the pressure interpolants is one order less than the degree for the velocities. The same kind of condition on polynomial spaces has to be imposed for spectral collocation and spectral element methods [10].

In the transient case, the pressure computation is in many applications a very time consuming part during the first time steps. This is due to the start up procedure of the problem which may be very stiff and induce very steep time variations. To overcome this difficulty, the time scheme and the pressure solver must be strongly stable [3].

4 Algebraic solvers

As soon as the time integrator is implicit or a Poisson equation is needed, one has to solve a large linear (or linearized) set of equations. Here, some algorithms made their breakthroughs and are unavoidable. To name a few of them, let us cite : multigrid [1], conjugate gradient, conjugate gradient square [14], GMRES [13], several preconditioners [16].

10

On new architectures, like massively parallel computers, iterative or semi-direct methods are very popular. Their generality and their stability of convergence are still in debate when they are compared with more robust techniques like multi-processor frontal method.

5 Other numerical considerations

5.1 On the ellipsoid case, outflow conditions have to be prescribed on the outer boundary. How do they affect upstream the computed flow field?

5.2 Is the parallelepipedic cavity problem well posed? Indeed, we have edge and corner singularities in terms of the velocity field. Locally, at these singular points, the divergence of the velocity field is blowing up and might hamper the convergence of the full problem.

5.3 What is the accuracy of the schemes in a long term integration over thousands of time steps? Are we sure that the numerical simulation is still close to the physics? Do we produce numerical artefacts, numerical chaos by dispersion errors, upwind schemes, artificial damping, personal recipes of numerical cooking (boundary conditions, tricks, "magic" numbers!)?

At the end of this short review, we may conclude that despite several decades of numerical experience, the solution of three-dimensional transient incompressible flows still raises open questions. Some of them have answers but not yet definite answers. The presence of complicated (internal or external) geometries will add more complexity to the numerics and more dynamical features to the physics. Therefore, the design of efficient and accurate Navier-Stokes solvers remains a challenging field for the numericists.

References

1. BRANDT, A., *Multi-level adaptive solutions to boundary value problems,* Math. Comput., 31, (1977), p.333-390.
2. CANUTO, C., HUSSAINI, M.Y., QUARTERONI, A., ZANG, Z.A., *Spectral Methods in Fluids Dynamics,* Springer, New-York, (1988).

3. DEVILLE, M., KLEISER, L., MONTIGNY-RANNOU, F., *Pressure and time treatment of Chebyshev spectral solution of a Stokes problem*, Int. J. Num. Meth. Fluids, 4, (1984), p. 1149-1163.

4. FLETCHER, C.A.J., *Computational Techniques for Fluid Dynamics*, vol. I, Springer, Berlin, (1991).

5. GIRAULT, V., RAVIART, P.A., *Finite Element Approximation of the Navier-Stokes Equations : Theory and Algorithms*, Springer, Berlin, (1986).

6. GOLUB, G.H., VAN LOAN, C.F., *Matrix Computations*, Johns Hopkins Univ. Press, Baltimore, (1983).

7. GRESHO, P.M., SANI, R.L., *On pressure boundary conditions for the incompressible Navier-Stokes Equations*, Int. J. Num. Meth. Fluids, 7, (1987), p.1111-1145.

8. GUNZBURGER, M.D., *Finite Element Methods for Viscous Incompressible Flows, A guide to Theory, Practice and Algorithms*, Academic Press, Boston, (1989).

9. LE, T.H., MORCHOISNE, Y., *Traitement de la pression en incompressible visqueux*, C.R. Acad. Sc. Paris, t. 312, Série II, (1991), p. 1071-1076.

10. MADAY, Y., PATERA, A.T., *Spectral Element Methods for the incompressible Navier-Stokes Equations, in State-of-The-Art Surveys on Computational Mechanics*, Eds : A.K. Noor, J.T. Oden, ASME, New-York, (1988), p. 71-143.

11. ORSZAG, S.A., ISRAELI, M., DEVILLE, M.O., *Boundary conditions for incompressible flows*, J. Sci. Comput., 1, (1986), p. 75-111.

12. PEYRET, R., TAYLOR, T.D., *Computational Methods for Fluid Flow*, Springer, New-York, (1983).

13. Y. SAAD, M-H. SCHULTZ, *GMRES : A generalized minimal residual algorithm for solving nonsymmetric linear systems*, SIAM J. Sci. Statist. Comput.,7, (1986), p. 856-869.

14. P. SONNEVELD, *CGS : a fast Lanczos-type solver for nonsymmetric linear systems*, SIAM J. Sci. Statist. Comput., 10, (1989), p. 36-52.

15. P.J. VAN DER HOUWEN, *Construction of integration formulas for initial value problems*, North-Holland, Amsterdam, (1977).

16. D.M. YOUNG, K.C. JEA, *Generalized Conjugate-Gradient acceleration of nonsymmetrizable iterative methods*, Lin. Alg. Appl., 34, (1980), p. 159-194.

PREDICTION OF THREE-DIMENSIONAL UNSTEADY LID-DRIVEN CAVITY FLOW

M. Arnal*, O. Lauer**, Ž. Lilek**, M. Perić**

*Lehrstuhl für Fluidmechanik, Technische Universität München, Arcisstr. 21,
  D-8000 München, FRG
**Lehrstuhl für Strömungsmechanik, Universität Erlangen, Cauerstrasse 4,
  D-8520 Erlangen, FRG

SUMMARY

This paper presents results of a numerical study of the unsteady, three-dimensional, lid-driven cavity flow at a Reynolds number of 3200. A finite-volume, multi-grid method was used in combination with a co-located variable arrangement to solve the governing equations. The central difference scheme is used for spatial discretization and two second-order schemes are employed for the time-discretization. The pressure and velocity fields are coupled using the SIMPLE algorithm.

It is shown that the flow is unsteady and non-periodic although the boundary conditions are steady in time. The spatial resolution was found to be much more important than the temporal resolution in determining the instantaneous flow field. The influence of imposing a symmetry boundary condition at the geometric symmetry plane of the cavity is considered. Results of the full-cavity simulation on a $32 \times 32 \times 96$ grid are presented and discussed.

INTRODUCTION

The lid-driven cavity flow is often used as a bench-mark case for testing numerical solution methods. Koseff *et al.* [1] performed a detailed experimental study of this flow in a cavity with a square cross-section and various widths, at relatively high Reynolds numbers. The authors observed that although the boundary conditions remained steady in time, the cavity flow itself was unsteady at $Re = 3200$. A series of Taylor-Görtler-like (TGL) vortices develop and move in time in the corner formed by the downstream and bottom walls. The same authors also performed a numerical simulation of the flow [2] and found good qualitative agreement with the experimental data. The present paper reports results of calculations for the same problem that were performed for the "GAMM Workshop on Numerical Simulation of 3-D Incompressible Unsteady Viscous Laminar Internal and/or External Flows".

The geometry and boundary conditions were those specified by Deville *et al.* [3] and are sketched in Figure 1. Calculations of the unsteady flow field were performed using two different time discretization schemes: Crank-Nicolson and three-time-level upwind (second order, fully implicit). The two cases simulated consisted of the flow in the full cavity geometry in one series of calculations and one symmetric half of the cavity in a second set. Both sets of calculations were carried out for a period of over 200 timescales, $T_c = L/U_b$, where U_b is the lid velocity, and L is the cavity width.

13

Special concern was given to checking the dependence of the solution on the choice of the temporal discretization scheme, the time-step size and the control-volume size (number of spatial grid points). The next sections describe briefly the numerical solution method used, followed by a presentation and discussion of the results. Finally, conclusions derived from the present study are summarized.

NUMERICAL SOLUTION METHOD

The problem under consideration is mathematically described by the conservation equations for mass and momentum. For a Newtonian fluid with constant density and viscosity the governing (Navier-Stokes) equations can be written as:

$$\frac{\partial(\rho U_i)}{\partial x_i} = 0 \tag{1}$$

$$\frac{\partial(\rho U_i)}{\partial t} + \frac{\partial}{\partial x_j}\left(\rho U_j U_i - \mu \frac{\partial U_i}{\partial x_j}\right) = -\frac{\partial P}{\partial x_i}. \tag{2}$$

Here ρ stands for fluid density, μ for its dynamic viscosity, U_i are the cartesian velocity components in the x_i-coordinate system and P stands for pressure (which includes gravity effects). On the boundaries of the solution domain the following boundary conditions apply:

- the tangential velocity component is equal to the wall velocity;

- the normal velocity component and its gradient in the normal direction are equal to zero (the latter resulting from the continuity equation leading to zero normal stress at the walls).

The solution domain is subdivided into a finite number of control volumes (CV) with a non-uniform mesh. The above equations are first formally integrated over a CV and time interval. After application of the Gauss theorem this leads to the following form:

$$F_e - F_w + F_n - F_s + F_t - F_b = Q. \tag{3}$$

In Eq. (3) F_e stands for the sum of the convection and diffusion fluxes through a given CV face 'e'. The source term Q includes the volume integral of the right hand side of Eq. (2) as well as the time derivative contribution. The fluxes are always evaluated with respect to surface vectors directed along the corresponding positive coordinate, hence the minus sign at faces 'w', 's' and 'b'. The discretization will be described for the face 'e'; other faces are treated analogously. Two difference schemes were used for the discretization in time: the two-time-level Crank-Nicolson scheme and a three-time-level fully implicit scheme. In the first case the fluxes and source term in the above equation are evaluated by averaging the fluxes at the two time levels, i.e.

$$F_e = 0.5(F^n + F^{n-1})_e \tag{4}$$

where the superscript n denotes the time level. In the second case the fluxes and source terms are all evaluated at the new time level. In the continuity equation, the flux F_e represents the mass flux through a surface:

$$F_e = \dot{m}_e = (\rho U_x \delta y \delta z)_e. \tag{5}$$

14

In the momentum equations, the momentum flux F_e consists of both a convection and a diffusion contribution. By applying the mean value theorem, the flux expression in this case can be written as:

$$F_e = \dot{m}_e U_{i,e} - \mu \left(\frac{\partial U_i}{\partial x} \delta y \delta z \right)_e .$$ (6)

At this stage the linearization and discretization of the flux expression is introduced. The mass flux $\dot{m}_e$ is treated as known (taken from the previous iteration), and the gradient at the CV face is approximated as (central-difference approximation):

$$\left(\frac{\partial U_i}{\partial x} \right)_e \approx \frac{U_{i,E} - U_{i,P}}{x_E - x_P} .$$ (7)

The cell-face value of the velocity, $U_{i,e}$, is here evaluated from a combination of the central (linear interpolation) and upwind approximations:

$$U_{i,e} = \gamma U_{i,e}^C + (1 - \gamma) U_{i,e}^U = U_{i,e}^U + \gamma (U_{i,e}^C - U_{i,e}^U)$$ (8)

The blending factor γ is chosen equal to unity (full central-difference approximation), unless the grid is too coarse, in which case some upwind contribution must be used to stabilize the solution procedure. The scheme is implemented in the so-called 'deferred correction' manner, i.e. only the upwind approximation is implicitly treated, and the correction term is added to the source.

The discretization of the pressure term leads to:

$$Q_{U_x}^p = -(P_e - P_w) \delta y \delta z .$$ (9)

In case of two-time-level schemes the time derivative is approximated as:

$$Q_U^t = -\frac{\rho \delta V}{\delta t} (U_{i,P}^n - U_{i,P}^{n-1}) .$$ (10)

In case of the three-time-level scheme, the time variation of the variable is approximated by a quadratic function. The variable values at the current and previous two time-levels are used to determine the coefficients of this function, and the derivative at the current time-level is obtained by differentiating the function, thus:

$$Q_U^t = -\frac{\rho \delta V}{2 \delta t} (3 U_{i,P}^n - 4 U_{i,P}^{n-1} + U_{i,P}^{n-2}) .$$ (11)

The scheme was implemented so that a first-order fully implicit (forward Euler) scheme could be used as an alternative by changing the value of the parameter, β:

$$Q_U^t = -\frac{\rho \delta V}{\delta t} \left[(U_{i,P}^n - U_{i,P}^{n-1}) + \beta \cdot \frac{U_{i,P}^n - 2 U_{i,P}^{n-1} + U_{i,P}^{n-2}}{2} \right] .$$ (12)

Note that when the parameter is set to $\beta = 0$ a forward Euler scheme is obtained. By setting the parameter to $\beta = 1$ the second-order three-time-level scheme is recovered. The parameter may also be set to an intermediate value ($0 < \beta < 1$) to obtain an approximation to the time-dependent term which is a linear combination of the two time-schemes. This is similar to the implementation of the deferred-correction scheme for the spatial discretization of the convection term.

15

When approximations for all fluxes and source terms are introduced in Eq. (3), the following algebraic equation in terms of the nodal variable values results:

$$A_P U_{i,P} = \sum_{nb} A_{nb} U_{i,nb} + Q_{U_i}^* \tag{13}$$

where the index 'nb' runs over the nearest neighbors of node P, i.e. E,W,N,S,T and B. The second quantity on the right-hand side (Q^*) contains the source terms and explicitly treated parts of the convection and time derivative terms. For the solution domain as a whole, a matrix equation with a seven-diagonal coefficient structure results. It is solved iteratively using the ILU-decomposition scheme after Stone [4].

Since the equations to be solved are non-linear and coupled, the above algebraic equation systems are not solved exactly for the given coefficient matrix; rather an iterative procedure is used. The velocity field obtained by solving the momentum equations is used to calculate the new mass fluxes. Since the co-located variable arrangement of variables is used, the velocities at CV faces, which are needed for the evaluation of mass fluxes, must be obtained by interpolation. Here the approach described by Perić et al. [5] is used. The mass fluxes calculated in this way do not satisfy the continuity equation and need be corrected. By requiring that the corrected mass fluxes satisfy the continuity equation, a pressure-correction equation is derived according to the well-known SIMPLE method [6]. The corrected mass fluxes and pressure are used to assemble the new coefficients and source terms of the momentum equations, and the whole process is repeated until convergence. The convergence criterion was that the sum of absolute residuals for all variables at the end of each time step must be below $10^{-5} - 10^{-6}$. In order to speed up the convergence rate for the number of iterations per time step, a multigrid procedure with V-cycles was employed, cf. Hortmann et al. [7]. Up to four grid levels were used in the present study.

Calculation times on the CRAY Y-MP 4/464 were approximately 18 CPU seconds per time-step for the simulation of the symmetric half of the cavity flow ($32 \times 32 \times 48$ grid-points) which required 2.9 MWords of core memory. For the full flow-field calculation the CPU time on the same computer was approximately 38 seconds per time-step for a grid of $32 \times 32 \times 96$ grid-points and 5.4 MWords of memory. Without the multigrid procedure the computing time would double for the equivalent calculations on a single grid. For finer grids (4 levels) an even more significant saving can be expected.

RESULTS AND DISCUSSION

Grid dependence analysis

The time discretization errors were estimated by performing calculations of the initial flow development up to a time $t = 5$. Time-step sizes of $\delta t = 0.2, \delta t = 0.1$ and $\delta t = 0.05$ were used with three different discretization schemes: first-order fully implicit (FI), second-order Crank-Nicolson (CN), and second-order fully implicit (SOFI). Solutions were obtained for the grid with $32 \times 32 \times 96$ CV using three-level multigrid procedure (the coarsest grid had $8 \times 8 \times 24$ CV). Variable values at several monitoring locations were recorded as a function of time. Figure 2 shows the result for the U-velocity component at one location obtained with the FI (a) and SOFI (b) schemes. Comparisons show that the solution with the SOFI scheme and $\delta t = 0.05$ are almost time-step independent, whereas the FI solution is still changing, with the change being towards the SOFI solution. Figure 3 shows the result obtained with all three schemes and the smallest time step, $\delta t = 0.05$. The two higher-order schemes give almost identical results, with little discrepancy among the three schemes. All results to be presented later were obtained either with the CN or SOFI schemes and a time-step size of $\delta t = 0.05$.

The spatial discretization errors were analysed for the calculations of one symmetric half of the cavity flow. This was done in order to calculate the flow on three different grids using the multigrid technique in each case. The finest grid consisted of $64 \times 64 \times 96$ CV in the half cavity geometry using a four-level multigrid procedure. Because of memory limitations it was not possible to simulate the full cavity flow with the equivalent fine-grid resolution. Coarser grid solutions consisted of $32 \times 32 \times 48$ CV and $16 \times 16 \times 24$ CV. The variations of the U-velocity component with time at two locations in the flow field on the three grids is shown in Figure 4. Clearly, there is a substantial change in the temporal development of the velocity as the grid is refined. The solution on the coarsest grid is not even qualitatively similar to the solutions on the two finer grids. Comparing Figures 2 and 4, it becomes evident that the spatial discretization error is orders of magnitude larger than the temporal discretization error. On the finest grid, new structures appear which are not seen on the coarser grids (Figure 4). Figure 5 shows centerline profiles of the U- and V- velocity components on the symmetry plane at time $t = 25$, predicted using the SOFI scheme on three different grids. A substantial grid dependence of the solution is demonstrated, leading to the conclusion that the spatial discretization errors on the $32 \times 32 \times 96$ CV grid are of the order of 10% of the lid velocity (U_b). The results to be presented and discussed later can be considered only as a qualitative representation of the flow studied. Calculations on finer grids, as soon as the computing resources become available, need to be performed in order to achieve higher accuracy of the solution.

Finally, a comparison of the two second-order time schemes was made. For a grid of $32 \times 32 \times 96$, calculations were performed with both CN and SOFI schemes up to a total elapsed time of 200 timescales. Results with both schemes remain in good agreement over the entire simulation time. Figure 6 shows a comparison of the W-velocity component at one monitoring location as a function of time. Although some differences exist at various times, these are small compared to both the temporal and spatial discretization errors due to the grid and time-step sizes. We note that for calculations performed on the $32 \times 32 \times 96$ CV grid with the SOFI time scheme, the maximum Courant number was 2 and the maximum cell Reynolds number was $Re_c = 130$.

Symmetry considerations

One of the questions to be answered in the study was whether the flow was truly symmetrical about the geometrical symmetry plane $z = 0$. Experimental observations indicate that the flow appeared to be symmetric [1], explaining why previous calculations were often done for one half of the cavity only [2]. Comparisons of the variable values on either side of the symmetry plane in our calculations of the full cavity showed that the flow does, in fact remain symmetrical at all times (at least for the given grid of $32 \times 32 \times 96$ CV). However, calculations performed for one half of the cavity, using the same grid ($32 \times 32 \times 48$ CV) and a symmetry boundary condition, lead to a somewhat surprising result. The flow development is nearly identical in both cases up to a time of

$t = 50$. Thereafter, the two solutions begin to depart, following different paths as shown in Figure 6 for the W-velocity component at one monitoring location. For the entire 200 timescales predicted, the temporal variation of the velocity and pressure at several other monitoring positions showed no evidence of converging to a regular, repeatable periodic behavior. Calculations of the half-cavity flow were continued for additional 200 timescales and still no periodic behavior developed. Note that since the flow is not periodic, it may well be that the mean and RMS velocities are still the same or similar in both cases.

The influence of the symmetry-plane condition is also demonstrated in Figure 7, which shows flow patterns in the plane $x = 0.267$ at time $t = 100$ for the whole (a) and half (b) cavity simulations. These figures were produced by following 'massless' particles in the direction of the velocity vector contained in the plane. Note that these are not particle paths in a 3-D flow, rather they represent velocity vector components in the plane, plotted so that they show the flow motion. All particles are followed for the same time period, so the length of each line is directly proportional to the magnitude of the corresponding velocity-vector component. Note that the strength, size and position of the TGL vortices is significantly different in the two plots. Since the time and space discretizations are identical in the two cases, the differences are due to the imposition of the symmetry-plane boundary condition.

Discussion

As mentioned above, neither of the two predicted flows (full and half solution domain) showed definite periodic behavior. However, a spectrum analysis of the time series at the various monitoring positions shows the presence of a dominant frequency at $f = 0.03332$ as well as several harmonics. The base frequency corresponds to a period of approximately 30 timescales. In order to determine whether additional fundamental frequencies are present and evaluate the flow structure responsible, it would be necessary to continue the calculations. However, in order to make this worthwhile the discretization errors would have to be significantly reduced.

Figures 8(a,b) illustrates again the dependence of the solution on the grid employed. The two calculations from which the figures are taken differ only in the distribution of the grid nodes in the z-direction. In Figure 8(a) a non-uniform grid is used ($\Delta z_{max}/\Delta z_{min} = 3.34$), while in Figure 8(b) a uniform grid is employed in the z-direction. We note in particular, that the eddy pair straddling the symmetry plane appears much weaker in 8(a). It is in this region that the resolution in the z-direction is poorest for the non-uniform grid case. In contrast, the uniform grid solution yields a velocity pair on the symmetry plane which is the strongest of those present at the given point in time.

Figures 7(a) and 8(a) illustrate the TGL vortices at the $x = 0.267$ plane at times $t = 100$ and $t = 200$. The vortices can be seen to vary in strength and meander in time. They first appear near the side walls ($z = -1.5, 1.5$) in the full-cavity calculations and induce additional vortices on the stationary lower wall as the flow develops. Note that for the half-cavity simulation the symmetry plane condition causes the vortex pair at the cavity mid-plane to develop more quickly. To see this, compare the vortex strengths on the symmetry plane in Figures 7(a,b). At all times the vortices appear only on the lower wall opposite the sliding lid and in the region of the downstream secondary eddy.

18

The concave curvature of this secondary eddy has been hypothesized to be the origin of the instability leading to the development of these TGL vortices. We note that, in the simulations, there are 9 pairs of vortices in the full cross-section. In the flow visualization studies at this Reynolds number, 9 pairs of vortices were also observed, giving qualitative support for the predicted results. At longer times the vortices become stronger in the region away from the side-wall. The vortices give a strong three-dimensional character to the flow field which would not be seen, of course, in two-dimensional simulations of the lid-driven cavity flow.

Figure 9(a,b) shows the flow pattern in two different z-planes at time $t = 200$. In both figures the primary recirculating region is observed, however the additional secondary eddies which occur in the corner of the cavity only appear in one figure or the other. This is an additional indication of the three-dimensional nature of the flow field.

Time development of the primary recirculating region is shown in Figure 10 which presents profiles of the U-velocity components in the geometric symmetry plane ($z = 0$). As the flow develops, a local maximum appears in both profiles which moves away from the bottom ($y = -1$) and upstream ($x = -1$) walls in time. This is an influence of the development and growth of the TGL vortex pair on the symmetry plane . In order to compare the predicted profiles with the time-averaged measurement profiles given in Perng and Street [8] the instantaneous profiles would have to be appropriately averaged.

Figure 11 shows wall-shear stresses on the side and bottom walls at $t = 200$. They indicate the three-dimensional vortex structure present in the vicinity of walls. The nine TGL vortex pairs can be clearly seen in Figure 11(b). The blank stripes indicate the boundaries of each vortex pair where a down-wash stagnation type flow occurs.

CONCLUSIONS

The results of the calculations presented here for the unsteady lid-driven cavity flow at $Re = 3200$ indicate good qualitative agreement with experimental observations. Grid dependence analysis shows that the spatial resolution has a much stronger influence on the flow development than the time-step size. Calculations with a refined grid need to be performed in order to make a quantitative analysis of the flow features.

REFERENCES

[1] KOSEFF, J.R., STREET, R.L.: 'The Lid-Driven Cavity Flow: A Synthesis of Qualitative and Quantitative Observations', *J. of Fluids Engineering*, **106**, 390-398, (1984).

[2] FREITAS, C.F., STREET, R.L., FINDIKAKIS, A.N., KOSEFF, J.R.: 'Numerical Simulation of Three-Dimensional Flow in a Cavity', *Int. J. Numer. Methods Fluids*, **5**, 561-575, (1985).

[3] DEVILLE, M., LÊ, T.H., MORCHOISNE, Y.: 'Testcase Specification of the GAMM Workshop on Numerical Simulation of 3-D Incompressible Unsteady Viscous Laminar Internal and/or External Flows', June, 12-14, 1991, Ecole Nationale Supérieure d'Arts et Métiers.

[4] STONE, H.L.: 'Iterative Solution of Implicit Approximations of Multi-Dimensional Partial Differential Equations', *SIAM J. Numer. Anal.*, **5**, 530-558 (1968).

[5] PERIC, M., KESSLER, R., SCHEUERER, G.: 'Comparison of Finite-Volume Numerical Methods with Staggered and Colocated Grids', *Comput. Fluids*, **16**, 389-403 (1988).

[6] PATANKAR, S.V., SPALDING, D.B.: 'A Calculation Procedure for Heat, Mass and Momentum Transfer in Three- Dimensional Parabolic Flows', *Int. J. Heat Mass Transfer*, **15**, 1787-1806 (1972).

[7] HORTMANN, M., PERIĆ, M., SCHEUERER, G.: 'Finite Volume Multigrid Prediction of Natural Convection: Benchmark Solutions', *Int. J. Numer. Methods Fluids*, **11**, 189-207 (1990).

[8] PERNG, C., STREET, R.L.: 'Three-Dimensional Unsteady Flow Simulations: Alternative Strategies for a Volume-Averaged Calculation', *Int. J. Numer. Methods Fluids*, **9**, 341-362 (1989).

FIGURES

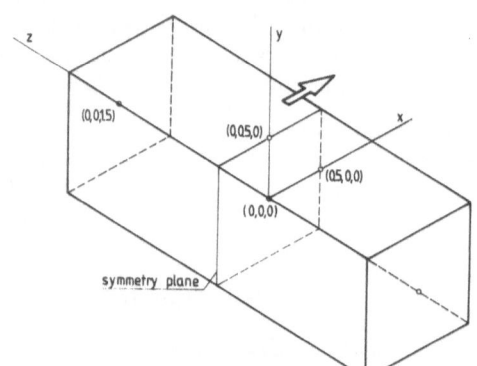

Fig. 1: Geometry of the solution domain

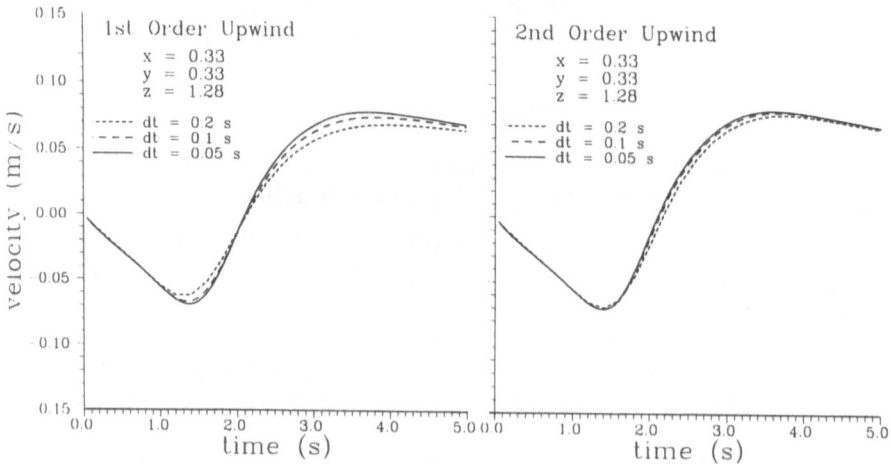

Fig. 2: Solution dependence on time-step size, two time schemes, 32 × 32 × 48 CV

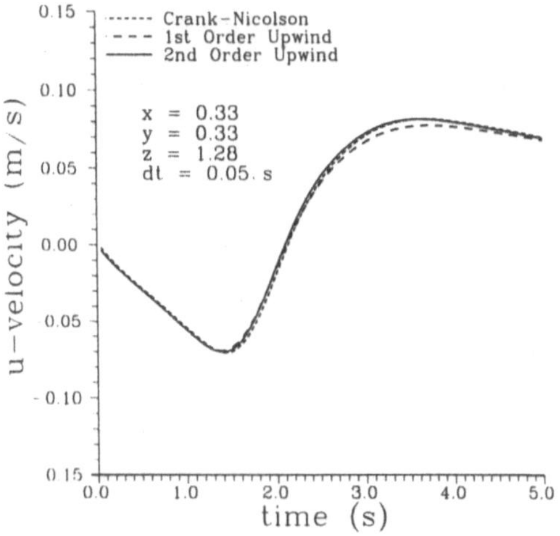

Fig. 3: Solution dependence on time differencing scheme, $32 \times 32 \times 48$ CV

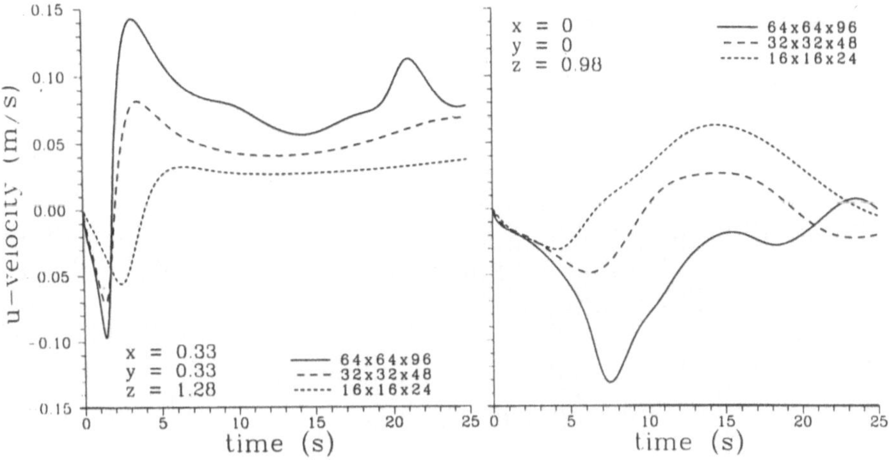

Fig. 4: Time development variation with spatial resolution, $\delta t = 0.05$

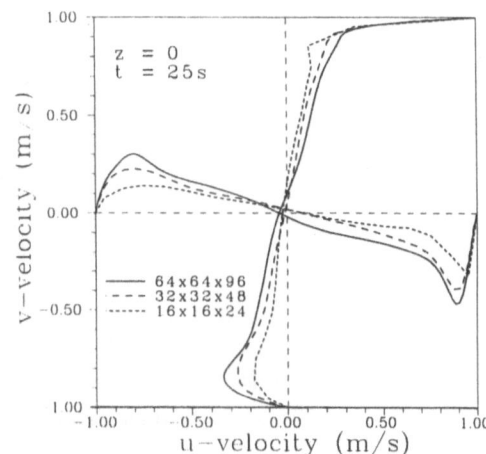

Fig. 5: Variation of velocity profiles
with spatial resolution

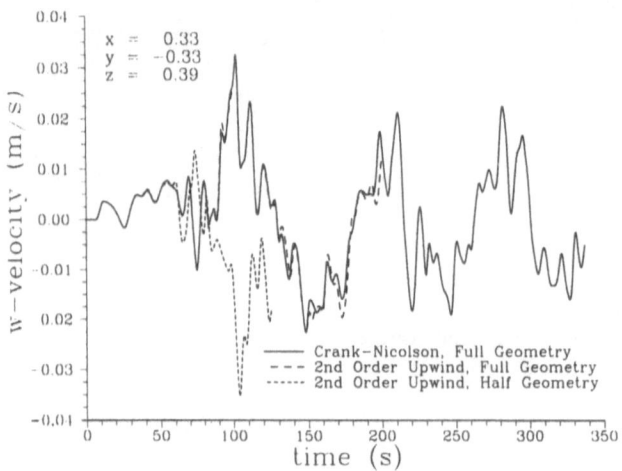

Fig. 6: Variation in temporal development with domain geometry at a monitoring location

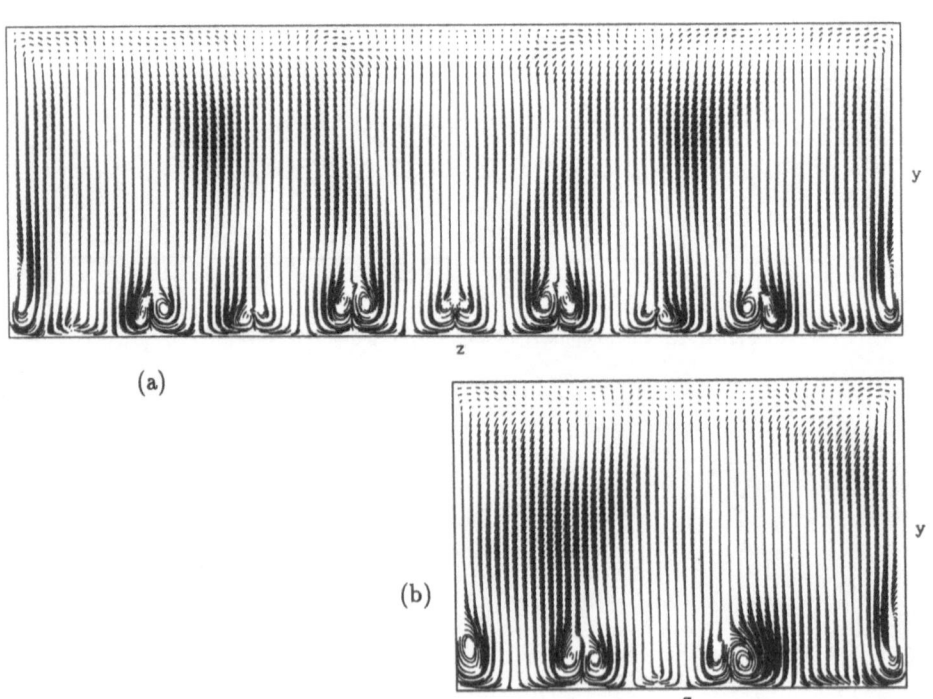

(a)

(b)

Fig. 7: Comparison of instantaneous flow patterns, $x = 0.267$, $t = 100$,
(a) full cavity, (b) half cavity

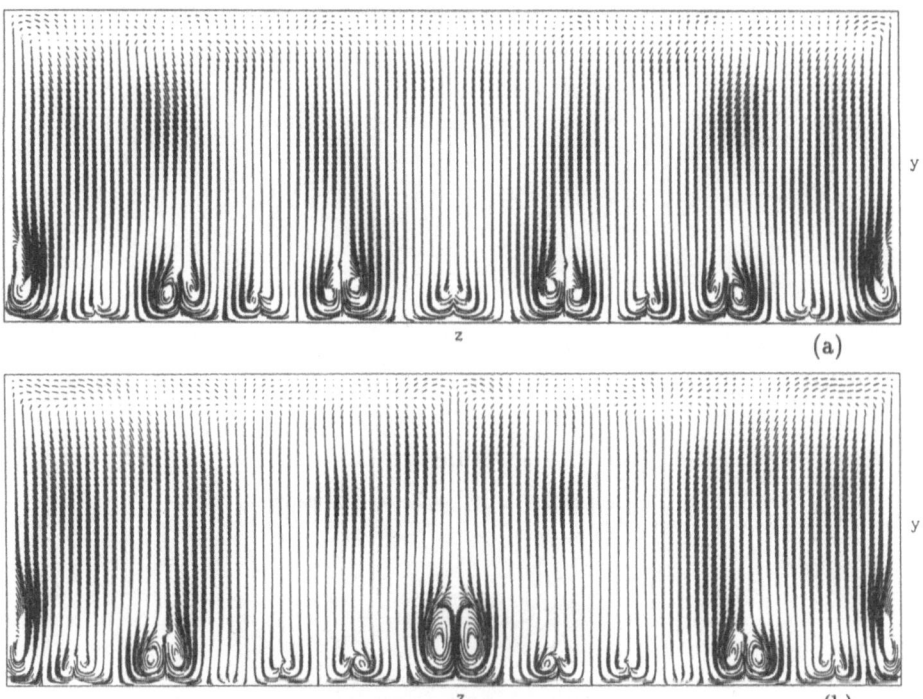

Fig. 8: Comparison of instantaneous flow patterns, $x = 0.267$, $t = 200$,
(a) non-uniform Δz, (b) uniform Δz

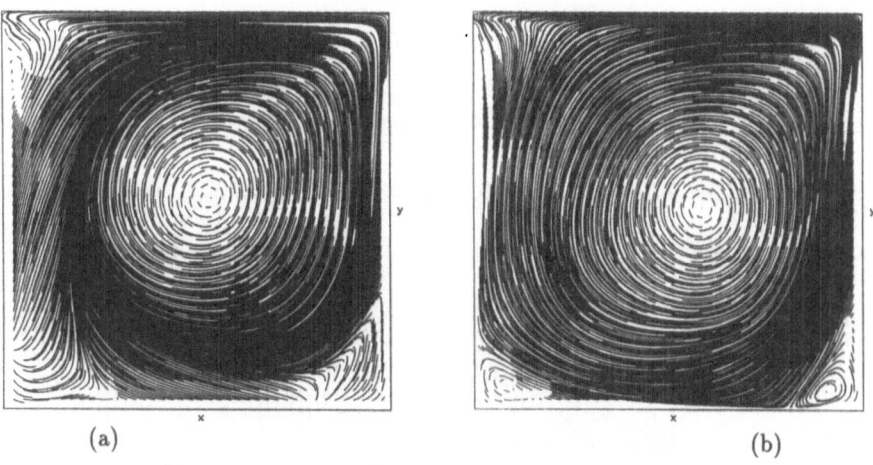

Fig. 9: Instantaneous cross-section flow patterns, $t = 200$,
(a) $z = 0.$, (b) $z = -1$.

23

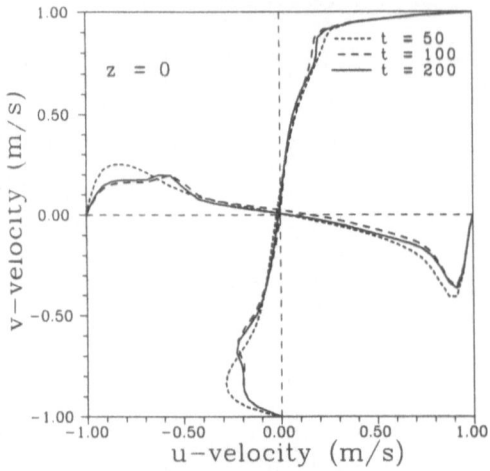

Fig. 10: Centerline velocity profiles at various times, $z = 0$.

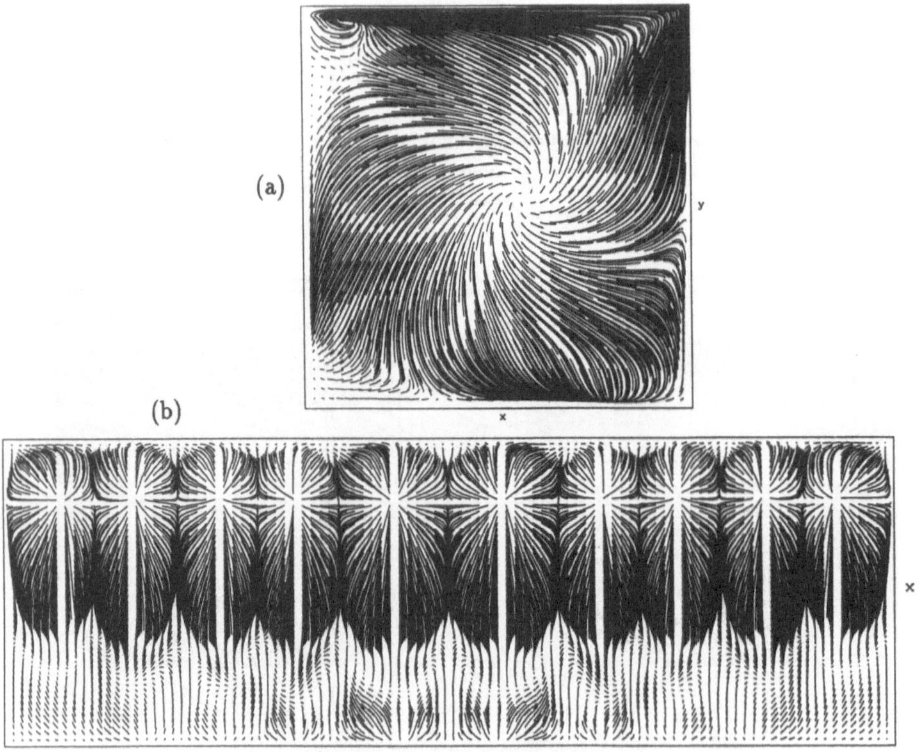

(a)

(b)

Fig. 11: Instantaneous wall shear stress patterns, $t = 200$,
(a) side wall, $z = 1.5$, (b) bottom wall, $y = -0.5$

DIRECT SIMULATION OF UNSTEADY FLOW

IN A THREE-DIMENSIONAL LID-DRIVEN CAVITY

B. Cantaloube, T.H. Lê

ONERA, B.P. 72, 92322 Châtillon Cedex, France.

SUMMARY

This paper deals with the direct simulation of the three-dimensional unsteady laminar flow of an incompressible viscous fluid in a lid-driven cavity at a Reynolds number equal to 3200. The dimensionless Navier-Stokes equations, using a velocity-pressure formulation, are written in conservative form. The discretization is based on a semi-implicit finite difference method with non-staggered variable arrangement. Central difference schemes are used for temporal and spatial discretization and are both second-order accurate. Pressure is obtained by solving a Poisson-like equation with a discrete divergence velocity minimization algorithm. Unsteady numerical results related to velocity, vorticity and pressure fields are given for three characteristic times after the upper surface of the cavity is impulsively started from rest.

INTRODUCTION

From the numerical viewpoint, the three-dimensional flows in a cavity serve as ideal prototype non-linear problems for testing numerical codes. Geometric simplicity and well defined flow structures make these flows very attractive as test cases for new numerical techniques and also provide benchmark solutions to evaluate differencing schemes and problem formulation.

Experimental and numerical 3-D results [1, 2] indicate the complexity of this flow with secondary eddies occuring in the longitudinal direction, well-known as the TGL (Taylor-G''ortler-like) vortices. These vortices develop in the flow as a result of the interaction of no-slip endwall effects and a concave streamline surface.

In the present study, three-dimensional calculations of the unsteady laminar flow within a cubic cavity were performed using the *PEGASE* code, developed at ONERA for the direct simulation of separated flows [3] and turbulent flows [4]. The driving force of the flow is a belt moving at the top of the cavity at a speed which produces a flow Reynolds number of 3200. Figure 1 displays the geometry and flow definitions.

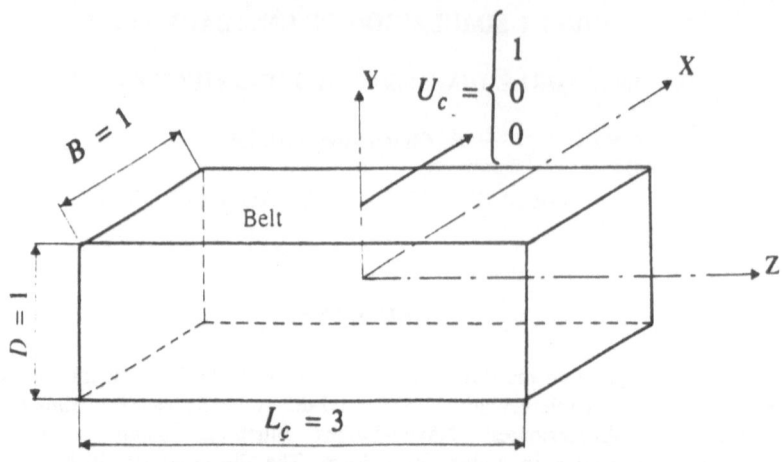

FIGURE 1 : Definition of the driven cavity

The box has a depth-to-width aspect ratio $\dfrac{D}{B}$ of 1 and a span-to-width aspect ratio $\dfrac{L_c}{B}$ of 3. At the top of the cavity a belt, impulsively started from rest, drags the adjacent fluid with a speed U_c which produces a Reynolds number of 3200 ($Re = \dfrac{U_c L_c}{\nu}$), ν is the kinematic viscosity. On the other walls non-slip velocity conditions prevail.

NUMERICAL METHOD

Governing equations

The equations governing the flow of an incompressible viscous fluid are the Navier-Stokes equations in a primitive variable formulation, written here in conservative form. The dimensionless equations are written as follows :

$$\frac{\partial U}{\partial t} + \nabla . (UU) + \nabla P = \frac{1}{Re} \nabla^2 U$$

$$\nabla . U = 0$$

where $U = U (x, y, z, t) = (u, v, w)^T$ is the three-dimensional velocity field, P is pressure field and Re is the Reynolds number. The characteristic scales are L_c for length and U_c for velocity.

Discretization procedure

The Navier-Stokes equations, including the continuity equation, are discretized on a non-staggered grid : velocity components and P are located at the same node which greatly alleviates the coding burden and enhances the applicability of the method. The numerical scheme is based on a method proposed in [5] with some improvements concerning the numerical scheme used for the discrete continuity equation. The convective terms are discretized in time by an explicit Adams-Bashforth scheme, while an implicit Crank-Nicolson method is used for the diffusive part to reduce stability constraints on the time step. The discretized equations are written as follows :

$$(1) \quad \frac{U^{n+1} - U^n}{\Delta t} + \frac{1}{2} (3H^n - H^{n-1}) + G (P^{n+\frac{1}{2}}) = \frac{1}{2\,Re} L (U^{n+1} + U^n)$$

$$(2) \quad D\,U^{n+1} = 0$$

where $G = (\delta_x , \delta_y , \delta_z)^T$ represents differencing along spatial directions, $H = (H_u, H_v, H_w)$ are the convective terms, L is the seven point discrete Laplacian operator and $D = (\delta_x , \delta_y , \delta_z)$ is the discrete divergence operator, where $\delta.$ is a second order difference operator.

Equation (1) is written in Δ - form as

$$(3) \quad (I - \frac{\Delta t}{2\,Re} L)\, \Delta\,U^{n+1} = -\Delta t\, [\frac{1}{2} (3H^n - H^{n-1}) - \frac{1}{Re} L (U^n) + G (P^{n+\frac{1}{2}})]$$

where I is the identity operator and $\Delta\,U^{n+1} = U^{n+1} - U^n$. For a given $P^{n+\frac{1}{2}}$ the previous equation is solved by introducing an implicit approximate factorization

$$(4) \quad (I - \frac{\Delta t}{2\,Re} L)\, \Delta\,U^{n+1} = A\, \Delta\,U^{n+1} + O(\Delta t^3)$$

where $A = A_x A_y A_z$ with $A_x = (I - \frac{\Delta t}{2\,Re} \frac{\delta^2}{\delta x^2})$ and similar formula for A_y and A_z.

The operator introduced in (4) is inverted by solving tridiagonal systems based on Lower-Upper decomposition.

Boundary conditions

The Navier-Stokes equations, for an incompressible flow, require boundary conditions on velocity, which are sufficient to allow the determination of both velocity and pressure. No-slip boundary conditions are assigned at the solid walls.

The problem specification is completed by prescribing initial conditions for the velocity field. Here the flow is at rest initially.

Pressure calculation

In the present work, a new scheme has been developed to solve the elliptic pressure equation, which yields an accurate solution of the mass conservation equation.

Applying operator $\mathbf{D}$ to equation (3), the following equation is obtained :

$$(5) \quad \mathbf{D}\, U^{n+1} = \mathbf{D}\, U^n - \Delta t\ \mathbf{D}\, \mathbf{A}^{-1}\, [\ \frac{1}{2}\, (\, 3\mathbf{H}^n - \mathbf{H}^{n-1}\,) - \frac{1}{Re}\, \mathbf{L}\, (\, U^n) + \mathbf{G}\, P^{n+\frac{1}{2}}\,].$$

Should the left hand side of (5) vanish which is equivalent to the following Poisson-like equation in $P^{n+\frac{1}{2}}$:

$$(6) \quad -\, \mathbf{D}\, \mathbf{A}^{-1}\, \mathbf{G}\, P^{n+\frac{1}{2}} = -\, \frac{\mathbf{D}\, U^n}{\Delta t} + \mathbf{D}\, \mathbf{A}^{-1}\, [\ \frac{1}{2}\, (\, 3\mathbf{H}^n - \mathbf{H}^{n-1}\,) - \frac{1}{Re}\, \mathbf{L}\, (\, U^n\,)\,],$$

then the mass conservation equation (2) will be satisfied.

The solution of (6) is obtained by means of eliminating the spurious modes and minimizing iteratively a discrete norm of the velocity divergence.

These spurious modes consist of the discretization modes due to the choice of a non-staggered grid (i.e : velocity and pressure are computed at the same nodes) and of the constant physical mode. The former (discretization) modes may be suppressed by lowering the number of degrees of freedom in pressure : pressure is computed only on inner nodes. Because of the latter mode, equation (6) must be minimized in a quotient space modulo constants, this implies that the residual $\mathbf{D}\, U^{n+1}$ may be a constant. This constant value is set to zero because the central difference scheme applied to the conservative divergence formulation is used with an odd number of points in the $x - y$ Dirichlet directions. This scheme satisfies a discrete compatibility condition , for more details see [6, 7]. First applications were done in [8] for a plane channel flow and in [9] for a rotating square pipe flow. As a result , because an iterative residual solver [10] is used , the constant pressure mode is automatically suppressed.

RESULTS

Numerical specifications

Calculations have been performed at ONERA on a CRAY YMP 4-128 computer.

A cartesian mesh has been used for the whole cavity with a constant spatial increment in each direction, no symmetry condition has been imposed. In x and y directions the mesh size is $\Delta x = \Delta y = .0125$ and in the z direction $\Delta z = .0130$, corresponding to 81x81x231 total number of points. A constant time step is used along computations and is equal to $\Delta t = .01125$; the maximum Courant number is .9 and the maximum cell Reynolds number is 40 .

The flow field has been computed up to $t = 200$ (the time scale t_c is equal to $\dfrac{U_c}{B}$), the corresponding total number of time steps is 17 778. The number of iterations required in the iterative solver, for the Poisson-like equation (6) with a tolerance on the divergence norm of $| D \ U^{n+1} |_\infty \leq 1.5 \ 10^{-4}$, depends on the unsteadiness of the flow. Indeed, during the stage ranging from $t = 0$ to $t = 8$ the average number of iterations is 20 and after it decreases to 5 . The average CPU time is 2.31 10^{-6} sec/node/time-step corresponding to a total cost of 17 hours.

Discussion of numerical results

Figure 2 displays the vorticity z-component and pressure contours in the symmetry plane ($z = 0$) . Negative and positive values are respectively represented by dashed and full lines. One can remark that the primary vortex occupying the central core of the flow is "fed' by a " vortex-sheet " issued from the upper downstream edge. Downstream, upstream and upper secondary eddies defined in [1] are well detected on figure 2 as well as on vorticity and pressure contours.

Figure 3 shows, in the symmetry plane ($x = 0$), the time evolution of the flow structures by means of vorticity y-component contours. The driving belt, located at the upper part of each figure, moves toward the reader. It is worthwhile noticing that till $t = 50$ (fig. 3-a) only 8 pairs of TGL vortices are evidenced; the central TGL vortex appears after $t = 50$ (fig. 3-b and 3-c) and 9 pairs of vortices are obtained as it is observed in experiments. The symmetry of the flow is observed for $t = 50$ and $t = 100$ (Fig 3-a and 3-b). At $t = 200$, the symmetry behaviour has disappeared and the TGL vortices seem to be more concentrated in the vicinity of the symmetry plane. Figure 4, displaying at $t = 200$ the pressure contours, emphasizes this loss of symmetry.

Velocity profiles for $t =50$, 100 and 200, plotted in the symmetry plane on center lines $x = 0$ and $y = 0$ (fig. 5 and 6), shows the unsteadiness of this flow. For comparison, a five minute-averages experimental results from [1] (from $t = 180$ to 203) are represented by full dots.

CONCLUSIONS

The *PEGASE* code with some improvements concerning the pressure calculation is used to compute the unsteady three-dimensional flow in the lid driven cavity at a Reynolds number of 3200 for which complicated structures are exhibited, the TGL vortices. Numerical results obtained assess the feasibility of the method for accurate simulations of such a flow.

In order to determine the stability of the flow structure without any doubt and to clarify the symmetry aspects, further calculations are required in terms of spacial accuracy and long-time integration.

REFERENCES

[1] Rhee H.S., Koseff J.R. and Street R.L. : " Flow visualization of a recirculating flow by rheoscopic liquid and liquid cristal techniques " Exp. Fluids, 2, 57-64 (1984).

[2] Freitas C.J. Street R.L. Findikakis A.N. and Koseff J.R. : "Numerical simulation of three-dimensional flow in a cavity " Int. J. Numer. Methods Fluids, 5, 561-575 (1985).

[3] Troff B., Lê T.H. and Loc T.P. : "A numerical method for the three-dimensional unsteady incompressible Navier-Stokes equations" J. Comput. and Appl. Math., Vol. 35, 1991, pp. 311-318.

[4] Dang-Tran K. and Veber G. : " Direct Numerical Simulation of Laminar and Turbulent Flows in a Diverging Channel" AIAA 22nd Fluid Dynamics, Plasma Dynamics & Laser Conference, Honolulu, Hawa''i (USA), June 24-26, 1991. AIAA Paper 91-1776. T.P. ONERA n^0 1991-64.

[5] Dang-Tran K. and Morchoisne Y. : " Numerical methods for direct simulation of turbulent shear flows " von Karman Institute Lectures Series 1989-3 on " Turbulent Shear flows ", Rhode-Saint-Genèse (Belgium). T.P. ONERA n^0 1989-12.

[6] Lê T.H. and Morchoisne Y. : " Traitement de la Pression en Incompressible Visqueux " C. R. Acad. Sci. Paris, t. 312. Série II, p. 1071-1076, 1991.

[7] Ryan J., Troff B., Lê T.H. and Morchoisne Y. : "Pressure calculation in incompressible flow" (to be submitted to Int. j. numer. methods fluids).

[8] Rida S. and Dang-Tran K. : "Direct Simulation of Turbulent Pulsed Plane Channel Flows" 8th International Symposium on "Turbulent Shear Flow", Munich (Germany), September 9-11, 1991. T.P. ONERA n^0 1991-120.

[9] Lê T.H., Ryan J. and Dang-Tran K. : " Direct simulation of incompressible, viscous flow through a rotating square channel " 9th GAMM Conference on Numerical Methods in Fluid Mechanics, Lausanne (Switzerland), September 25-27, 1991.

[10] Ryan J., Lê T.H. and Morchoisne Y. : " Panel code solvers " Proceedings of the Seventh GAMM-Conference on Numerical Methods in Fluid Mechanics, Notes on Numerical Fluid Mechanics, Vol. 20, 1988, Michel Deville (Ed.), Vieweg, pp. 335-342. T.P. ONERA n^0 1987-139.

[11] Perng C.Y. and Street R.L. "Three-dimensional unsteady flow simulations : Alternative strategies for a volume-averaged calculation " Int. J. Numer. Methods fluids, 9, 341-362 (1989).

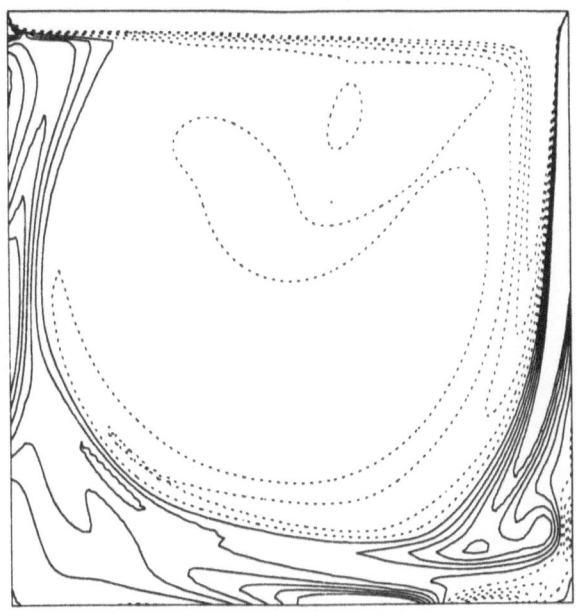

(2a)   Contours of z-component of the vorticity   ($-5. \leq \Omega_Z \leq 5.$)

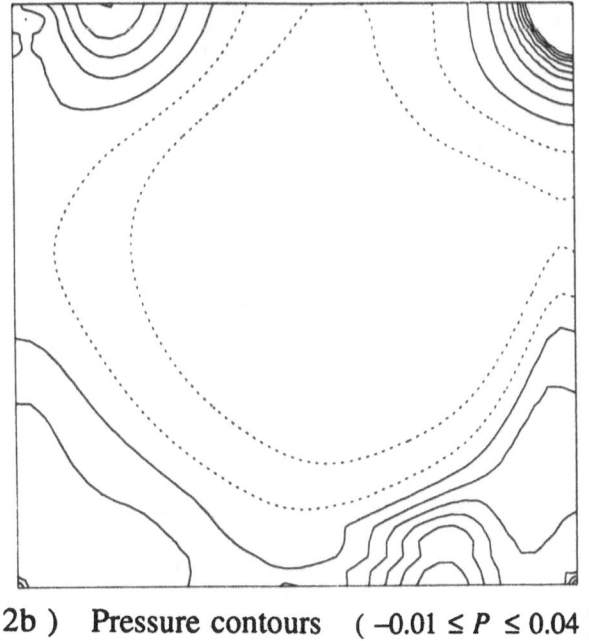

(2b)   Pressure contours   ($-0.01 \leq P \leq 0.04$)

Figure 2 : Flow structures at t=200 in the symmetry plane

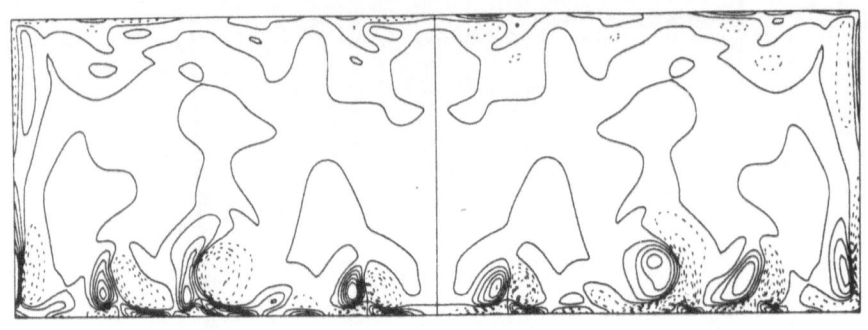

(3a) $t = 50$

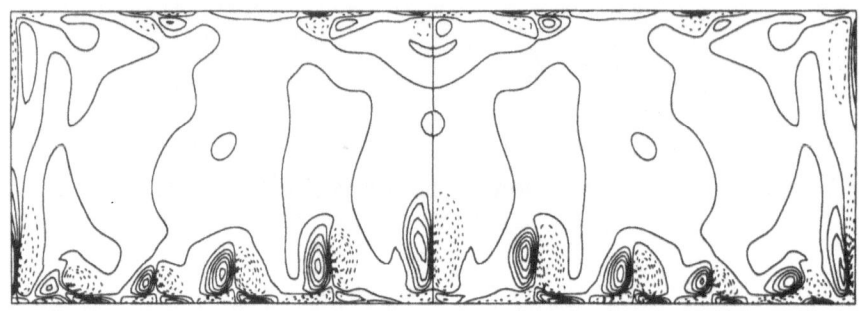

(3b) $t = 100$

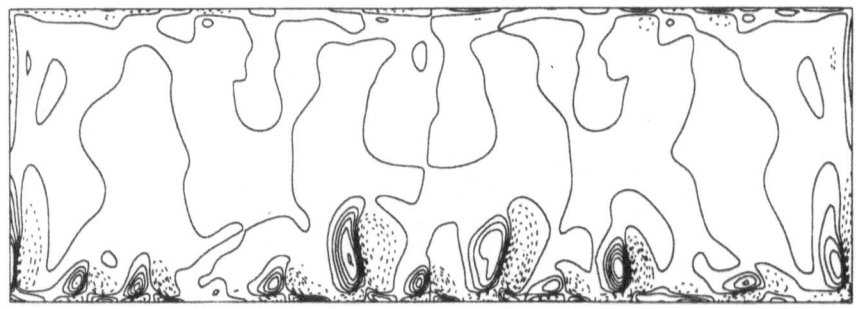

(3c) $t = 200$

Figure 3 : Contours of x-component of the vorticity
in x = 0 plane ($-5. \leq \Omega_x \leq 5.$)

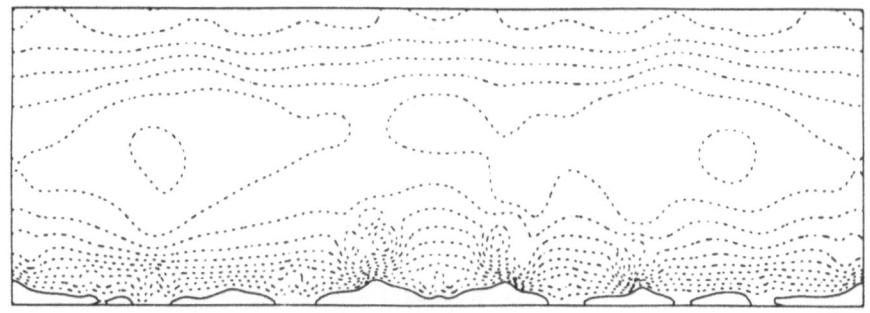

Figure 4 : Pressure contours at t=200
in x = 0 plane ($-0.02 \leq P \leq 0.$)

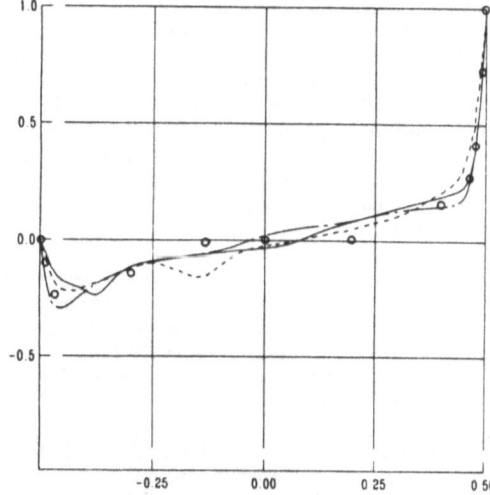

●: Experimental results

− − − : $t = 50$

−·− : $t = 100$

——— : $t = 200$

Figure 5 : u-Velocity profiles in z = 0 plane
at x=0.0

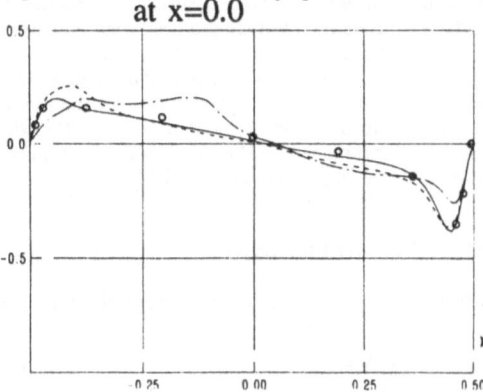

Figure 6 : v-Velocity profiles in z = 0 plane
at y=0.0

A Fully Implicit and Fully Coupled Approach for the Simulation of Three-Dimensional Unsteady Incompressible Flows

PARIS GAMM WORKSHOP. JUNE 1991

G.B. Deng, J. Piquet, P. Queutey & M. Visonneau

CFD Group, LHN-UA 1217-CNRS
1, Rue de la Noë, Nantes, FRANCE

1 INTRODUCTION

The flow of a viscous fluid in a three dimensional cavity driven by its sliding upper wall is an interesting benchmark test problem because of its simple geometry and boundary conditions. In contrast with its twodimensional counterpart which is the most often used test case, only a limited number of steady flow calculations have been performed in the past on the (cubic) 1:1:1 case, usually for low Reynolds numbers (< 1000) and a low grid resolution, with finite difference methods [1] to [12], finite element methods [13] or spectral methods [14]. For Reynolds numbers higher than 1000, a very few numerical solutions are available [6][13][15][16] since the flow behavior exhibits significant transverse motions, like Taylor-Görtler-like (TGL) vortices, endwall vortices and, for a given Reynolds number, stronger unsteady effects than in the 2D case. This situation is unfortunate since significant experiments can be realised only for the threedimensional case while Hopf bifurcation, periodicity, period doubling and numerical transition to the turbulence have been observed numerically only in two dimensions, due to the lack of computer ressources.

In the following, a numerical simulation of the flow in a 3:1:1 (spanwise aspect ratio - SAR- equal to 3) driven cavity has been performed. This case is one of those considered in the experiments [17][18][19] and in the (still coarse grid) calculations [15]. Short time survival Taylor-type toroidal vortices are observed soon after the lid is impulsively started. TGL vortices are present (and predicted in [13][15][16]) in the fully established state. Finally, turbulence occurs for Reynolds numbers in the range from 6000 to 8000.

The following prediction of the lid driven square cavity flow with SAR of 3:1 (fig.1) and Re=3200 is based on recent improvements of the discretization and resolution approaches. Following [20], a new interpolation approach, called the Consistent Physical Interpolation (CPI), is applied to the conservative finite difference formulation. It provides a second order accurate scheme on a non-staggered grid and it allows an efficient fully coupled iterative solution approach using a diagonal block preconditionned conjugate gradient method based on the so-called CGSTAB procedure [21]. The simulation of three dimensional incompressible unsteady flows is made possible with reasonable (second order) time and spatial accuracy on presently available surpercomputers. However, due to the still high cpu cost of the method, the unsteady evolution has been studied only up to an adimensional time t = 100 with a 64*64*64 grid, using a symmetry condition (in the plane z = 0). Preliminary calculations on a coarser grid indicates no spontaneous symmetry breaking, at least for t ≤ 100.

2 THE NUMERICAL APPROACH

2.1. *Master Equations.* The governing equations are the unsteady Navier-Stokes Equations :

$$\nabla . U = 0 \; ; \frac{\partial U}{\partial t} + \nabla . UU = - \nabla p + \frac{1}{Re} \nabla^2 U \tag{1}$$

with the following boundary conditions :

$$U = 0 \text{ for } x=\pm 1/2 \; ; \; y=-1/2 \; ; \; z=\pm 1/2) \tag{2a}$$

$$U = i \text{ for } y=+1/2 \text{ where } i \text{ is the unit vector along the x axis.} \tag{2b}$$

and initial condition :

$$U(x,y,z)|_{t=0} = 0. \tag{2c}$$

The equations (1) with boundary and initial conditions (2) are solved in the primitive variable formulation on a cell-centered collocated grid. A three level second order backward Euler scheme is chosen for time discretization. Spatial derivatives are treated implicitly by using central difference scheme on a cell-centered collocated grid. However, first order derivatives of the velocity field require intermediate values (fluxes) to be defined at faces of the control volume. For example (fig.2),

$$\frac{\partial u}{\partial x} \approx \frac{U_{i+1/2,j,k} - U_{i-1/2,j,k}}{\Delta x}. \tag{3}$$

The success of the numerical approach depends entirely on the way the numerical flux $U_{i\pm1/2}$ is interpolated.

2.2. *The Closure Problem.* In order to preserve the accuracy of the second order centered difference scheme, the interpolation approach needs to be at least second order accurate. Conventional closure methods, based on Taylor series expansions, lead usually to spatial red-black oscillations associated to spurious pressure modes of the solution. The closure [20], which circumvents such difficulties is obtained through the use of an interpolation technique based on the momentum equations. The present three-dimensional simulation uses this interpolation with a cell-centered collocated grid. Its principle is best explained if the following 2D steady case is considered (fig.2). Introducing the seven-point stencil and the standard cardinal notation of cell boundary points E-NE-NW-W-SW-SE-C, the following relation can be obtained by the discretization of the u-momentum equation:

$$C_C U_C = \Sigma\, C_{nb} U_{nb} + \frac{P_E - P_W}{2h_x} \tag{4}$$

where nb= (E,NE,NW,W,SW,SE), from which a closure formula relating U_C to its neighbouring independent variables U_{nb}, V_{nb} (appearing in the summation) *and* P_{nb} can be easily deduced. Any discretization technique can be used to obtain (4), however the following two requirements have to be fulfilled. (i) the formulation must be consistent ; (ii) the coefficient C_C must not be small for stability reasons (then the truncation error will not be amplified). Due to the special grid stencil, the solution is not straightforward. Convective terms can be treated easily by using a first order upwind scheme :

$$u\frac{\partial u}{\partial x} \approx U_C \frac{U_C - U_W}{h_x} \text{ for } U_C \geq 0 \text{ ; } = U_C \frac{U_E - U_C}{h_x} \text{ for } U_C < 0 \tag{5}$$

$$v\frac{\partial u}{\partial y} \approx V_C \frac{2U_C - U_{SW} - U_{SE}}{2h_y} \text{ for } V_C \geq 0 \text{ ; } V_C \frac{U_{NW} + U_{NE} - 2U_C}{2h_y} \text{ for } V_C < 0). \tag{6}$$

While the pressure gradient is centered, diffusive terms are treated using the approximation (7), despite its general lack of diagonal dominance :

$$\nabla^2 u \approx \frac{1}{2}\left(\frac{U_{NE} - 2U_E + U_{SE}}{h_y^2} + \frac{U_{NW} - 2U_W + U_{SW}}{h_y^2}\right) + \frac{U_E - 2U_C + U_W}{h_x^2}. \tag{7}$$

Although a first order scheme is used, the resulting closure formulation (8) :

$$U_C = f(U_{nb}, P_{nb}) + \frac{O(h_x, h_y)}{\frac{|U_C|}{h_x} + \frac{|V_C|}{h_y} + \frac{2}{Re}\frac{1}{h_x^2}} \tag{8}$$

is at least of second order accurate. (8) also shows that a consistent discretization is needed to get a second order closure. Due to the importance of the consistency, this closure approach is named as "Consistent Physical Interpolation" (CPI) approach. The extension to three-dimensional unsteady flows simulation is straightforward.

2.3. *The Resolution Approach*. With a fully implicit scheme, a non linear system needs to be solved at each time step t^n. The difficulty of obtaining a solution of this system lies in the lack of a pressure time derivative term in the continuity equation. The most commonly used methods follow the so-called seggregated or Poisson equation approach in which the pressure-velocity coupling is solved iteratively, updating sussessively the velocity variables in the momentum equations and the pressure in a pressure equation, in such a way that solenoidality is satisfied at convergence, the updating procedure being handled by the well known "SIMPLE", "UZAWA", "SIMPLER" or "PISO" methods. Here, following [22], the global system of unknowns is solved in a fully coupled way so that velocity and pressure fields are updated simultaneously. However, a collocative grid approach is used. This makes the eventual implementation of block-iterative methods (with higher rates of convergence) easier than with the staggered grid method [22].

2.4. *Congugate Gradient Acceleration*. For elliptic problems, the convergence rate of classical basis iterative methods, like Jacobi or Gauss-Seidel methods, is approximately proportional to N^{-2}, N being the number of unknowns per direction. When applied to the solution of the linearized algebraic system resulting from the discretization of the unsteady or steady incompressible Navier-Stokes equations, the behaviour of a block-iterative fully coupled method is about the same (For seggregated methods, the convergence rate is small with respect to N^{-2}, see [23]). As a result, the number of iterations increases dramatically with the number of grid points. For this reason, convergence acceleration, like multigrid or conjugate gradient, is needed, both for steady and for unsteady flow calculations, especially in the three-dimensional case. When a basis iterative method is used as a smoother or a preconditionner, the convergence rate of a multigrid method (independent of N) is higher than that of a conjugate gradient method (approximately proportional to N^{-1} according to our experience). In spite of its lower convergent rate, the CG method has been used for the following reasons: (i) It is independent of grid configuration and easier to implement than MG methods. (ii) the efficiency of a multigrid method depends strongly on the control parameters or process, such as interpolation / restriction operators, construction of coarse grid operators, choice of the smoother and of its relaxation factor, boundary condition treatment, influence of grid aspect ratio, etc. Unlike the multigrid method, no control parameter is needed for the CG method. (iii) Also our experience indicates that the computational effort for the CG method is of about the same order as that of the MG method when the number of grid points is not too high (say less than 100 by direction).

For a non-symmetric linear system, the CGS algorithm [24] is found to be very efficient. However, when applied to a non-linear problem with conventional (non-Newton) linearization, the computational effort is prohibitive because the CGS method does not satisfy a minimization property so that a non-monotone decrease of the error is found, even for an unsteady simulation where the time change remains small. A variant of CGS, the so-called "CGStab" algorithm [21] is used since it allows a smoother convergence and a better efficiency for time dependent problems. The efficiency of a CG method depends on the condition number of the matrix problem. Linearized algebraic equations resulting from the discretization of the Navier-Stokes equations are usually ill-conditionned, especially when large grid aspect ratios are used. Preconditionning is then needed. For the lid driven cavity problem, the block diagonal matrix is found to be the most efficient preconditionner on vector computers.

2.5. *Defect Correction Approach*. Although better results could presumably be obtained with momentum equations discretized in conservative form, resulting linearized algebraic equations are usually difficult to solve. Problems arise in the case of high Reynolds number where flow recirculation exists. Then the flow may have opposite convective directions one some control volumes, as shown in fig. 3. With a large cell Reynolds number, the diagonal term may lose its dominance. As a result, the condition number of the matrix increases considerably. The resulting linearized system fails to be solved even with the Block-ILU preconditionner. An alternative defect-correction approach is adopted here, since the diagonal dominance of the momentum equations can always be ensured, using a positive scheme to discretize the convective form of the transport equations. Let $L_1U = g$ be the discrete problem to be solved. A defect-correction approach is written as:

$$L_2U^{(n+1)} = L_2U^{(n)} + g - L_1U^{(n)} \qquad\qquad n=1,2,3,\ldots \qquad\qquad (9a)$$

or

$$L_2\Delta U^{(n+1)} = R^{(n)} \text{ with } U^{(n+1)} = U^{(n)} + \Delta U^{(n+1)}, R^{(n)} = g - L_1U^{(n)} \qquad (9b)$$

if L_2 is a linear operator. L_1 is the "explicit" operator which gives the accuracy, L_2 being an implicit operator which provides convergence. L_2 is selected as being :

$$\nabla.U^{(n+1)} = 0 ; \frac{3U^{(n+1)}-4U^{(n)}+U^{(n-1)}}{2\Delta t} + [U^{(n)}.\nabla]\ U^{(n+1)} = -\nabla p^{(n+1)} + \frac{1}{Re}\ \nabla^2U^{(n+1)} \qquad (10)$$

where the convective transport equations are discretized using the multi-exponential scheme [25] or any other positive scheme. Also, the discretization resulting from the CPI approach is retained for the continuity equation as the explicit operator. As a result, the mass conservation is ensured when the solution is updated. No convergence problem has been found so far with this defect-correction approach.

3 NUMERICAL RESULTS ; DISCUSSION

3.1.*Grid spacing and time step.* As shown in [17][18][19], the lid driven cavity flow is unsteady at Re = 3200. "Taylor" vortices are observed in the starting phase, and "Taylor-Görtler-like" vortices at later time. A lack of grid resolution may lead to spurious numerical results and failure to predict such physical phenomena. It is therefore necessary to fix the required grid spacing. Numerical results obtained with a coarse grid on the cubic lid driven cavity (33*33*17 in the half domain) at Re=400 are first compared with the pseudospectral 21*21*11 mode solution of [14]. Corresponding 2D results are also presented with the fine grid solution [27] (fig.4). Compared with results [14], discrepancies are observed mainly on the V profile near the downstream wall even with a higher grid solution. It is believed that this is due to the regularized cavity used in [14]. We can see that the grid resolution required for the 3D simulation is almost the same as that needed for the 2D calculations. We can thus determine the appropriate grid resolution by comparing the 2D results with fine grid solutions [26]. With a 65*65 uniform grid, good results are obtained for Re=1000 (fig.5), and reasonable results for Re=3200 (fig.6). A marginal improvement is obtained using a non-uniform grid. Other preliminary calculations with a 50*50*60 grid over the whole domain up to T=100 shows that the flow remains symmetric. For this reason, *a 65*65*65 uniform grid is used to perform the 3D simulation on the half domain with symmetric boundary conditions.* Steady velocity profiles obtained with the above grid solution at lower Reynolds number Re=100,400 and 1000 are presented in the symmetry plane on the appropriate centerlines (fig.7). Some characteristic values are given in Table 1.

Table 1

Re	U_{min}	V_{min}	V_{max}	D_D	B_D	B_U	D_U
100	-.211	-.246	.172	.138	.132	.073	.074
400	-.300	-.441	.279	.286	.243	.121	.110
1000	-.331	-.477	.310	.325	.275	.208	.177

An inadequate time step may also lead to erroneous results. Time step limitations may either result from numerical (stability) limitations or from physical limitations which cannot be avoided with an improved numerical approach. The use of an explicit or semi-implicit approach usually imposes a (CFL type) time step limitation depending on the grid spacing. Such a limitation may be prohibitive when a fine grid is used. The implicit fully-coupled approach does not suffer from such a limitation. Nevertheless, a time step limitation still exists due to the resolution approach for the non linear equations. The defect-correction approach and the preconditionned CG solver allow a large time step simulation. However it has been found that a converged time solution could not be obtained with a time step greater than 2.

This practical limitation is believed to be due to physical reasons since experiments [19] indicate that the developed cavity flow is quasi-periodic with a period about T=2 (15 seconds).

Due to the solenoidality constraint, the total computational cost increases considerably when the time step is decreased. Convergence histories of the linearized equation for four different time steps (Δt=100, 10, 1., 0.1) at time t = 100 with the 50*50*60 grid are compared in fig.8.a for "CGS" and in fig.8b for "CGSTAB". It can be seen that "CGSTAB" is more efficient than "CGS" when used for time-dependent problems. The CPU time to solve the linearized system as well as the required number of non-linear iterations (not shown) are reduced when the time step is decreased, but not enough to make the total computational time independent of the used time step. Consequently, for a fully coupled implicit approach, the most economical time step is the maximum time step allowing a good description of the physical events. As indicated previously, the lid driven cavity flow at Re=3200 tends to become periodic with a period about T=2. *So the time step is fixed at Δt=0.25 for t > 50 while the initial phase (t < 50) is described with Δt = .5*, in order to save CPU time. The computation has been performed on the CRAY2 supercomputer. The reduction factor for the residuals of the non-linear equations is 10^3 while the residuals for the corresponding linearized equations are reduced by more than four orders of magnitude. The CPU time per point per time step is about 2.4 ms for Δt=0.5 and 1.6 ms for Δt=0.25. The storage cost of the fully coupled implicit method is about 45 MWords.

3.3. *Results*. To illustrate the flow evolution, four velocity profiles v(0, -0.4, z, t), u(0, -0.4, z, t), u(0, y, 0, t) and v(x, 0, 0, t) are presented in fig.9. A peak in the profile v(0, -0.4, z,t) indicates the presence of a pair of TGL vortices. The time evolution, in the symmetry plane z = 0, of the primary vortex centre (x_C, y_C), identified by the location of the minimum velocity modulus, is presented in figs.8. The birth of the primary vortex follows quickly the impulsive start of the lid. Because of the end wall, a corner vortex is formed. For, say, t < 10, the influence of the corner vortex is limited only to a small region close to the end wall. Unlike the experimental results where toroidal Taylor vortices were observed during the first 30 seconds (t < 5), no significant three dimensional effect occurs except near the end wall. A simulation performed up to t = 2 with a smaller time step (Δt=0.1) gives similar results. Disagreement between simulation and experiment is thus unlikely due to unsufficient time or spatial lack of resolution.The centre of the primary vortex is displaced nearly along the diagonal line during this period (fig.8b).

The corner vortex induces evenly distributed spanwise TGL vortices. The first pair of TGL vortices was observed around t = 15 (fig.9c), It induces a second pair about t = 25 (fig.9e). The important 3D effect propagates from the endwall, with the TGL vortices, from the end wall, and a pair of vortices centered at the symmetry plane appears at t ≈ 36 (fig.9h). Flow becomes quickly unstable in the whole cavity. The first distinct nine pairs TGL vortices pattern is observed about t = 47 (figure 9i). Afterwards, the flow is characterized by the strong interaction between the primary vortex, the TGL vortices and the corner vortices. Although nine pairs of TGL vortices are observed at t = 50 and t = 100, only eight pairs, intermittently observed during the simulation (fig.10), are known to occur predominantly in the experiments [17].

Although the existence of TGL vortices, as well as the main flow pattern, are expected to depend only on the Reynolds number, the end wall plays an important role in the main flow development. The main flow is already fully developed at t ≈ 25 (fig.9e), the TGL vortices do not cover the whole domain until t ≈ 47, as a result of the interaction between the first TGL vortices propagated from the end wall and the symmetry vortices formed at t ≈ 36 (fig.9h) due to the zero flux condition.

Numerical mean velocity profiles at the symmetry plane, for 50 < t < 100, are compared in fig.11 with experimental results taken in the fully established state with about five minutes sample averages (about t = 43). The u-velocity profile agrees well with experiments. The v velocity profile, however, is strongly influenced by the TGL vortices and

38

it does not agree as well. This is probably due to the sample of the numerical simulation data rather than to the grid resolution. Fig.12 compares a calculated time trace of the velocity v at point x = .333, y = -.397, z = 0, with an experimental sample taken at x = .333, y = -.4, z = 0, but from different time origins. TGL vortices influence the flow in the symmetry plane with a period which is well predicted. High frequency time changes are observed in the computation and in the experiments near the minimun point, indicating an unstable state. The numerical sample is smoother, however, probably because of the smoothing effect of the three level time discretization.

4 CONCLUSION

The new "consistent" physical interpolation, inspired by [20], allows the use of a non staggered grid without producing spurious pressure modes, while enforcing numerical stability and second order accuracy. Reliable numerical results have been obtained here with a moderate grid resolution. Due to the influence of the end wall, the TGL vortices develop in the present case without any numerical superimposed perturbation. However, the simulation has been performed only one Reynolds number and during a short time (about ten "physical" minutes). This is clearly not sufficient to understand the physical phenomenon in all its aspects, and particularly, its long time behaviour. The simulation has to be performed over longer times and also for a larger range of Reynolds numbers.

Acknowledgements. Cpu on Cray 2 has been provided by the Scientific Committee of CCVR (CNRS) (project 2409). Cpu on VP200 has been provided by DS/SPI (CNRS).

5 REFERENCES

[1] Takami, H. & Kuwahara, K. "Numerical Study of Threedimensional flow in a cubic cavity", *J. Phys. Soc. Japan* **37**, 1695-1698 (1974).
[2] De Vahl Davis, G. & Mallinson, G.D. "An Evaluation of Upwind and Central Difference Approximations by a Study of Recirculating Flow", *Comp. Fluids* **4**, 29-43 (1976).
[3] Goda, T. "A Multistep Technique with Implicit Difference Schemes for Calculating Two or Three-Dimensional Cavity Flows", *J. Comp. Phys.* **30**, 76-95 (1979).
[4] Dennis, S.C.R., Ingham, D.B. & Cook, P.N. "Finite Difference Methods for Calculating Steady Incompressible Flows in three dimensions" *J. Comp. Phys.* **33**, 325-339 (1979).
[5] Cazalbou, J.B., Braza, M. & Ha Minh, H. "A Numerical Method for Computing Threedimensional Navier-Stokes Equations apllied to cubic Cavity Flows with Heat TRansfer", *Num. Meth. Lam. Turbulent Flows* **4**, 786-797 (1983) Pineridge Press.
[6] Koseff, J.R., Street, R.L., Gresho, P.M., Upson, C.D., Humphrey, J.A.C. & To, W.M. "A Threedimensional lid-driven cavity flow : experiment and simulation", *Num. Meth. Lam. Turbulent Flows* **4**, 564-581 (1983) Pineridge Press.
[7] Kim, J. & Moin, P. "Application of a Fractional Step Method to Incompressible Navier-Stokes Equations", *J. Comp. Phys.* **59**, 308-323 (1985).
[8] Vanka, S.P. "A Calculation Procedure for Threedimensional Steady Recirculating Flows using Multigrid Methods", *Comp. Meth. in Appl. Mech. and Eng.* **55**, 321-328 (1986).
[9] Osswald, G.A., Ghia, U. & Ghia, K.N. "A Direct Algorithm for Solution of Incompressible three-dimensional Unsteady Navier-Stokes equations", *AIAA Paper 87-1139.*
[10] Perng, C.Y. & Street, R.L. "Improved Numerical Codes for Solving Threedimensional Unsteady Flows", *Num. Meth. Laminar and Turbulent Flows* **5**, 12-22 (1987) Pineridge Press.
[11] Hwang, D.P. & Huynh, H.T. "A Finite Difference Scheme for Threedimensional Steady Laminar Flow", *Num. Meth. Laminar and Turbulent Flows* **5**, 244-260 (1987) Pineridge Press.
[12] Orth, A. & Schönung, B. "Calculation of Threedimensional laminar Flows with complex boundaries using a Multigrid Method", *Notes Num. Fluid Mech.* **23** ; Wesseling, P. Ed. (1990) Vieweg Verlag.
[13] Kawai, H., Kato, Y., Sawada, T. & Tanahashi, T. "GSMAX-FEM for Incompressible Viscous Flow Ana í ysis (A modified GSMAC method)", *JSME Int. J.* **33-1**, 17-30 (1990).

[14] Ku, H.C., Hirsh, R.S. & Taylor, T.D. "A Pseudospectral method for Solution of three-dimensional unsteady Navier-Stokes Equations", *J. Comp. Phys.* **70**, 439-462 (1987).

[15] Freitas, C.F., Street, R.L., Findikakis, A.N. & Koseff, J.R. "Numerical Simulation of Three-dimensional Flow in a Cavity", *Int. J. Num. Meth. Fluids* **5**, 561-575 (1985).

[16] Iwatsu, R., Ishii, K., Kawamura, T., Kuwahara, K. & Hyun, J.M. "Numerical Simulation of Three-dimensional flow structure in a driven cavity", *Fluid Dynamics Research* **5**, 173-189 (1989).

[17] Koseff, J.R. & Street, R.L. "Visualization Studies of a Shear Driven Three-Dimensional Recirculating Flow", *J. Fluids Eng.* **106**, 21-29 (1984).

[18] Koseff, J.R. & Street, R.L. "On End Wall Effects in a Lid-Driven Cavity Flow", *J. Fluids Eng.* **106**, 385-389 (1984).

[19] Koseff, J.R. & Street, R.L. "The Lid-Driven Cavity Flow : A Synthesis of Qualitative and Quantitative Observations", *J. Fluids Eng.* **106**, 390-398 (1984).

[20] Schneider, G.E. & Raw, M.J. "Control Volume Finite-Element Method for Heat Transfer and Fluid Flow using Collocated Variables - 1. Computational Procedure", *Num. Heat Transf.* **11**, 363-390 (1987).

[21] Van Der Vorst, H..A. & Sonneveld, P. "CGSTAB : a more smoothly converging variant of CG-S", preprint (1990).

[22] Vanka, S.P. "Block Implicit Multigrid Solution of Navier-Stokes Equations in Primitive Variables", *J. Comp. Phys.* **65**, 138-158 (1986).

[23] Deng, G.B., Piquet, J. & Visonneau, M. "Viscous Flow Computations using a Fully-coupled Technique", Proc. 3rd. Osaka Int. Coll. Num. Ship Viscous Flow, Tanaka, I., Suzuki, T. & Himeno, Y. Eds. (1991).

[24] Sonneveld, P. "CGS : A fast Lanczos-type Solver for Nonsymmetric linear Systems", *SIAM J. Sci. Statist. Comput.* **10**, 36-52 (1989).

[25] Deng, G.B. "Résolution des Equations de Navier-Stokes tridimensionnelles. Application au calcul d'un raccord plaque plane-aile", PhD Thesis, Univ. Nantes (1989).

[26] Ghia, U., Ghia, K.N. & Shin, C.T. "High-Re Solution for Incompressible Flow Using the Navier-Stokes Equations and Multigrid Method", *J. Comp. Phys.* **48**, 387-411 (1982).

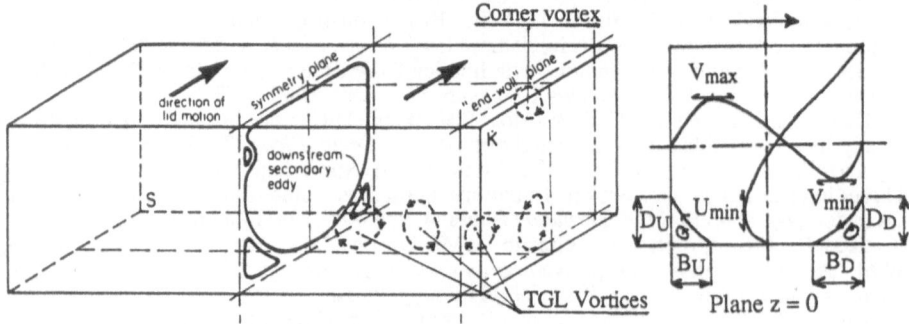

Figure 1. Definitions for lid-driven cavity flow

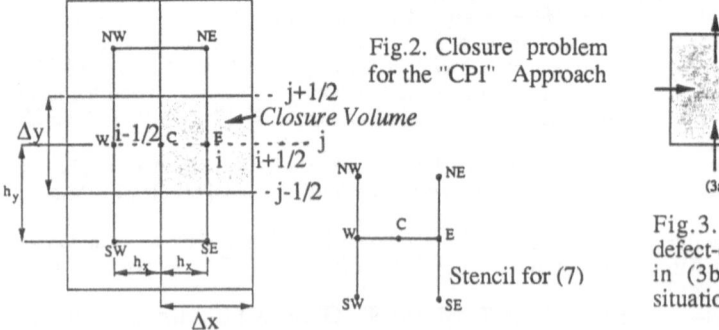

Fig.2. Closure problem for the "CPI" Approach

Stencil for (7)

Fig.3. Need for the defect-correction approach in (3b). (3a), "regular" situation.

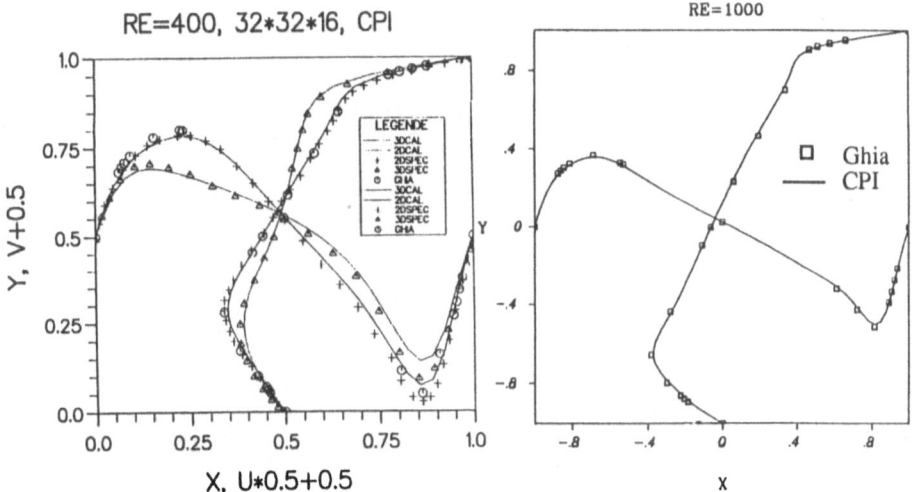

Figure 4. Cubic Cavity. Velocity Profiles for Re = 400 on vertical and horizontal centerlines. O, 2D Results [26]. +, 2D results [14]. Δ, 3D results [14]. — present "CPI" results.

Figure 5a. Velocity profiles in a square cavity ; Re = 1000. O, 2D Results [26]. —, present "CPI" results with a 65*65 grid.

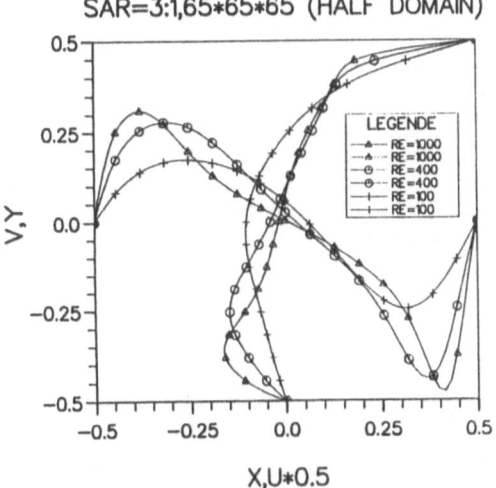

Figure 6. SAR 3:1:1 3D cavity. Velocity profiles on the vertical and horizontal centerlines for several values of Re.

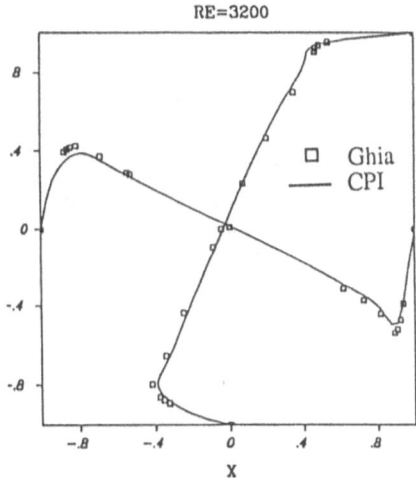

Figure 5b. Velocity profiles in a square cavity ; Re = 3200. O, 2D Results [26]. —, present "CPI" results with a 65*65 grid.

41

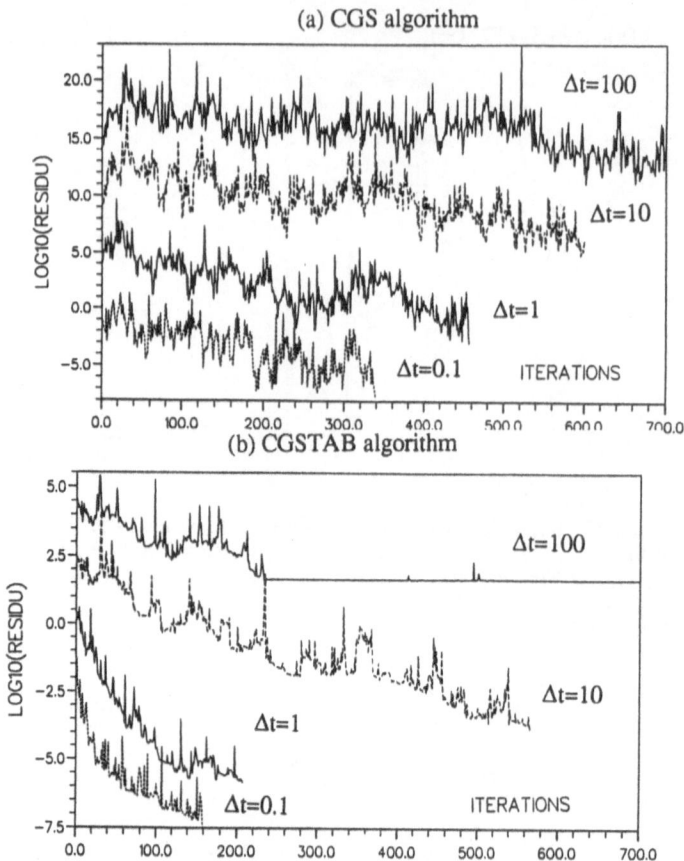

Figure 7. SAR 3:1:1 3D cavity. Re = 3200 ; t = 100. Convergence history of CG methods. Preconditioning with a block-diagonal matrix. 50*50*60 grid.

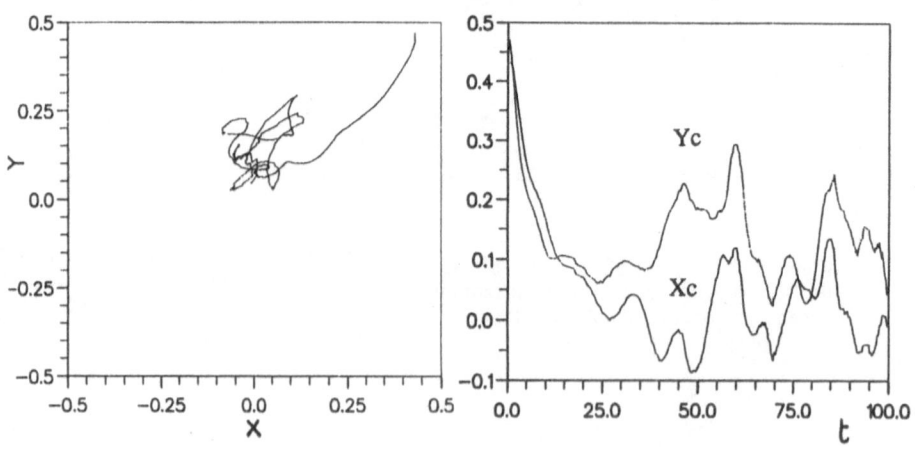

Figure 8. Impulsively started lid-driven SAR3:1:1 cavity. Re = 3200.
Evolution with time, in the symmetry plane z = 0, of the primary vortex centre.

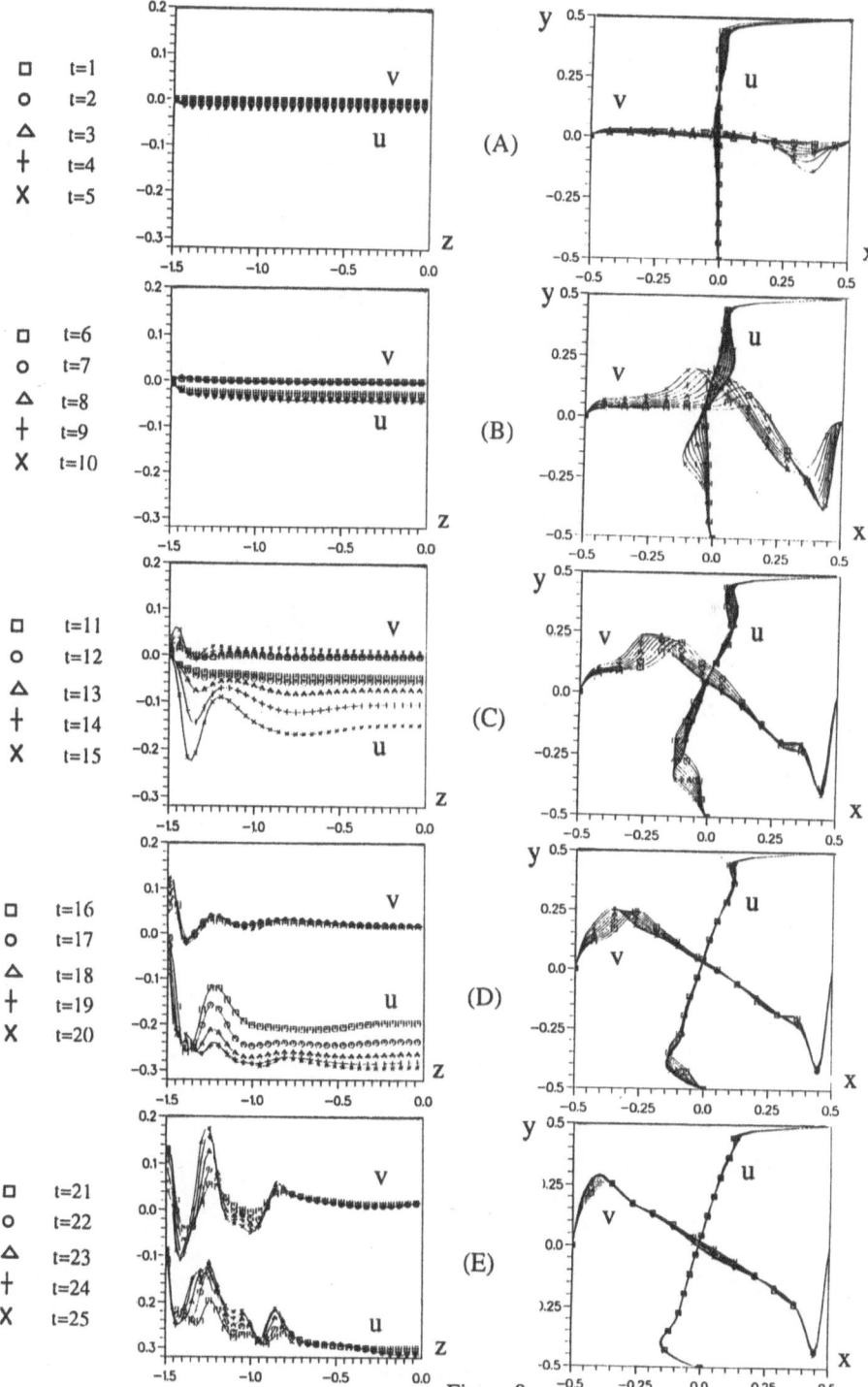

Figure 9.

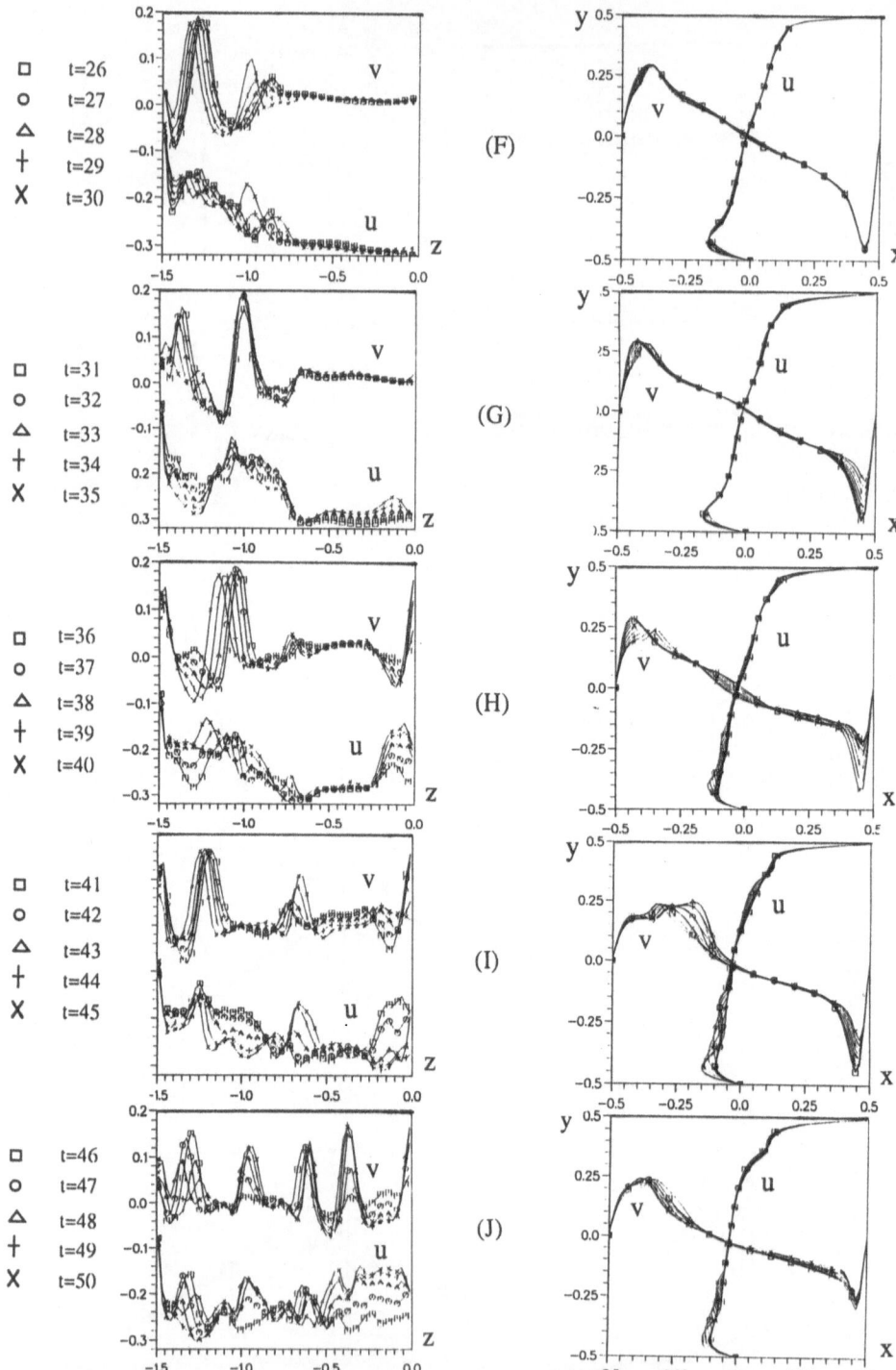

Figure 9. SAR 3:1:1 3D cavity. Re = 3200. Evolution with time of selected velocity profiles. For each left rectangle, upper curves, v(0, -0.4, z, t) ; lower curves, u(0, -0.4, z, t). For each. right rectangle, horizontal curves, v(x, 0, 0, t) ; vertical curves, u(0, y, 0, t).

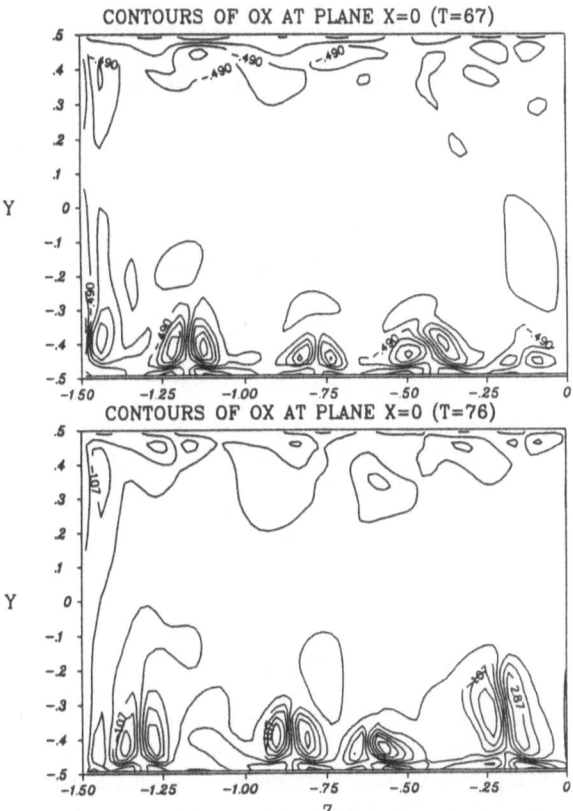

Figure 10. SAR 3:1:1 3D cavity. Re = 3200. Pattern with eight pairs of TGL vortices.
Contours of the x-component of the vorticity field in plane x = 0.
(left, endwall ; right, symmetry plane)

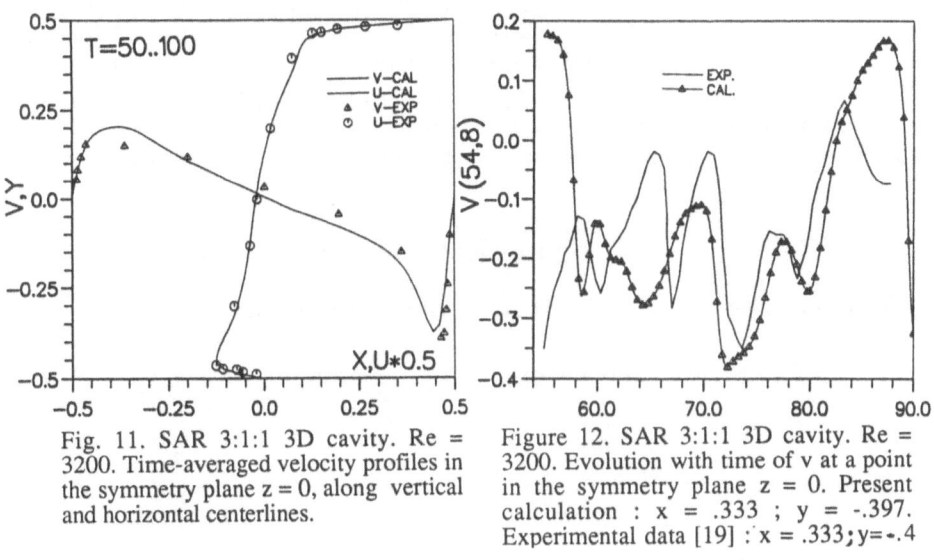

Fig. 11. SAR 3:1:1 3D cavity. Re = 3200. Time-averaged velocity profiles in the symmetry plane z = 0, along vertical and horizontal centerlines.

Figure 12. SAR 3:1:1 3D cavity. Re = 3200. Evolution with time of v at a point in the symmetry plane z = 0. Present calculation : x = .333 ; y = -.397. Experimental data [19] : x = .333 ; y = -.4

45

Numerical Simulation of a Three-Dimensional Lid-driven Cavity Flow

Pier Giorgio Esposito

I.N.S.E.A.N.
Via di Vallerano 139 – 00128 Rome, Italy

Summary

In the present work a multigrid technique for the solution of elliptic equations is applied to the Kim & Moin fractional step method [1]. The Navier–Stokes equations are discretized on a fixed step grid. This approach allows to obtain a very simple and fast code; furthermore there is no need of special smoothing operators to increase the convergence rate of the multigrid solver. The simulation has been performed on a workstation with a grid of about 750.000 cells.

1 Introduction

The investigation of a three-dimensional lid-driven cavity flow can be a useful benchmark for numerical schemes, because this flow shows well-defined structures combined with a very simple geometry. The geometric simplicity allows to evaluate the efficiency of the numerical scheme as it is, without the introduction of complex transformations which could affect the overall accuracy. From a physical point of view, this flow exhibits a primary stationary vortex, which is disturbed by the appearance of secondary three-dimensional and unsteady structures known as Taylor-Görtler-like (TGL) vortices. Even for spanwise aspect ratios (SAR) as high as 3:1 the flow is three-dimensional, because the presence of the TGL vortices forbids a two dimensional flow. For these reasons, lid-driven cavity flows have been investigated experimentally during past years. Koseff & Street [2,3] performed systematic LDA measurements in cavities with varying SAR; they found that the global three-dimensionality arises from end-wall no-slip boundary conditions, which results also in a weaker primary circulation with respect to the two-dimensional flow.

2 Numerical scheme

A primitive variables formulation of the Navier–Stokes equations has been chosen:

$$\frac{\partial \mathbf{u}}{\partial t} + \nabla \cdot (\mathbf{uu}) = -\nabla p + \frac{1}{Re}\nabla^2 \mathbf{u}, \tag{1}$$

where $\mathbf{u} = \mathbf{u}(x, y, z, t) = (u, v, w)$ is the three-dimensional velocity field which has to verify the continuity equation

$$\nabla \cdot \mathbf{u} = 0. \tag{2}$$

Equations (1) and (2) are discretized on a a staggered grid; velocity components and pressure are located as shown in Figure 2.

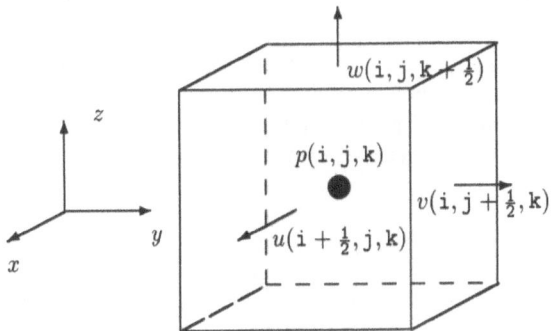

Figure 1: Sketch showing variables collocation

The numerical scheme is based on a fractional step method (see Kim & Moin [1]). The convective terms are discretized in time by an explicit Adams–Bashforth scheme, while an implicit Crank–Nicholson method is used for the diffusive part to reduce stability constraints on the time step. The solution is given by the two-step procedure:

$$\frac{\hat{\mathbf{u}} - \mathbf{u}^n}{\Delta t} + \frac{1}{2}[3\mathbf{H}^n - \mathbf{H}^{n-1}] = -\mathbf{G}(p^n) + \frac{1}{2Re}\mathbf{L}(\mathbf{u}^n + \hat{\mathbf{u}}), \tag{3}$$

$$\frac{\mathbf{u}^{n+1} - \hat{\mathbf{u}}}{\Delta t} = -\frac{1}{2}\mathbf{G}(\Phi), \tag{4}$$

where $\mathbf{G} = (\frac{\delta}{\delta x}, \frac{\delta}{\delta y}, \frac{\delta}{\delta z})$ represents central differencing along spatial directions, $\mathbf{H} = (H_u, H_v, H_w)$ are the convective terms, and $\mathbf{L}$ is the seven point discrete Laplace operator.

Equation (3) can be rewritten in Δ-form as

$$\left(\mathbf{I} - \frac{\Delta t}{2Re}\mathbf{L}\right)\Delta\hat{\mathbf{u}} = \Delta t\left\{-\frac{1}{2}[3\mathbf{H}^n - \mathbf{H}^{n-1}] - \mathbf{G}(p^n) + \frac{1}{Re}\mathbf{L}\mathbf{u}^n\right\}, \tag{5}$$

where $\mathbf{I}$ is the identity operator and $\Delta\hat{\mathbf{u}} = \hat{\mathbf{u}} - \mathbf{u}^n$. The previous equation is solved by introducing an implicit approximate factorization

$$\left(\mathbf{I} - \frac{\Delta t}{2Re}\mathbf{L}\right)\Delta\hat{\mathbf{u}} = \left(\mathbf{I} - \frac{\Delta t}{2Re}\frac{\delta^2}{\delta x^2}\right)\left(\mathbf{I} - \frac{\Delta t}{2Re}\frac{\delta^2}{\delta y^2}\right)\left(\mathbf{I} - \frac{\Delta t}{2Re}\frac{\delta^2}{\delta z^2}\right)\Delta\hat{\mathbf{u}} + \mathcal{O}(\Delta t^3). \tag{6}$$

The operator introduced in (6) can be inverted by solving tridiagonal systems.

The new pressure field can be calculated by requiring that the combination of (3) and (4) recast the discretization of the Navier–Stokes equations at the step $n + \frac{1}{2}$. It is easy to obtain

$$p^{n+1} = p^n + \frac{1}{2}\left[\Phi - \frac{\Delta t}{2Re}\mathbf{L}\Phi\right]. \tag{7}$$

The first step (3) provides the intermediate velocity field $\hat{\mathbf{u}}$; the scalar field Φ is introduced to enforce the solenoidality constraint on the field $\mathbf{u}^{n+1}$. By taking the discrete divergence D of Equation (4), an elliptic equation for Φ is obtained

$$\mathbf{L}\Phi = \frac{2}{\Delta t}D(\hat{\mathbf{u}}). \tag{8}$$

Since a great amount of computer time is spent to solve (8) a multigrid technique has been introduced to speed up the computation, as it will be described in Sec. 3.

The convective terms are written is a conservative form which is proved to preserve global momentum and energy in absence of time discretization errors and helps in keeping calculation stable for such a long-term integration [4]. A second order centered discretization is used. As an example the convective term of the x-momentum equation can be written as

$$H_u(\mathbf{i}+\tfrac{1}{2},\mathbf{j},\mathbf{k}) = \frac{1}{\Delta x}[\tilde{u}^2(\mathbf{i}+1,\mathbf{j},\mathbf{k}) - \tilde{u}^2(\mathbf{i},\mathbf{j},\mathbf{k})]$$

$$+ \frac{1}{\Delta y}[\tilde{u}(\mathbf{i}+\tfrac{1}{2},\mathbf{j}+\tfrac{1}{2},\mathbf{k})\tilde{v}(\mathbf{i}+\tfrac{1}{2},\mathbf{j}+\tfrac{1}{2},\mathbf{k}) - \tilde{u}(\mathbf{i}+\tfrac{1}{2},\mathbf{j}-\tfrac{1}{2},\mathbf{k})\tilde{v}(\mathbf{i}+\tfrac{1}{2},\mathbf{j}-\tfrac{1}{2},\mathbf{k})]$$

$$+ \frac{1}{\Delta z}[\tilde{u}(\mathbf{i}+\tfrac{1}{2},\mathbf{j},\mathbf{k}+\tfrac{1}{2})\tilde{w}(\mathbf{i}+\tfrac{1}{2},\mathbf{j},\mathbf{k}+\tfrac{1}{2}) - \tilde{u}(\mathbf{i}+\tfrac{1}{2},\mathbf{j},\mathbf{k}-\tfrac{1}{2})\tilde{w}(\mathbf{i}+\tfrac{1}{2},\mathbf{j},\mathbf{k}-\tfrac{1}{2})], \tag{9}$$

where $(\tilde{u},\tilde{v},\tilde{w})$ are obtained by averaging neighbouring values.

2.1 Boundary conditions

No-slip boundary conditions are assigned at solid boundaries. Near the boundaries, shear stresses are evaluated with a second order accuracy by a three point non-centered formula; as an example, at the bottom side wall $\frac{\partial u}{\partial z}$ is written as

$$\left.\frac{\partial u}{\partial z}\right|_{wall} \simeq \frac{-8u_{wall}(\mathbf{i}+\tfrac{1}{2},\mathbf{j}) + 9u(\mathbf{i}+\tfrac{1}{2},\mathbf{j},1) - u(\mathbf{i}+\tfrac{1}{2},\mathbf{j},2)}{3\Delta z}, \tag{10}$$

thus the part of diffusion term with derivatives normal to the wall becomes

$$\frac{\partial^2 u}{\partial z^2}(\mathbf{i}+\tfrac{1}{2},\mathbf{j},1) \simeq \frac{1}{\Delta z}\left(\frac{\delta u}{\delta z}(\mathbf{i}+\tfrac{1}{2},\mathbf{j},1) - \left.\frac{\partial u}{\partial z}\right|_{wall}\right) =$$

$$\frac{8u_{wall}(\mathbf{i}+\tfrac{1}{2},\mathbf{j}) - 12u(\mathbf{i}+\tfrac{1}{2},\mathbf{j},1) + 4u(\mathbf{i}+\tfrac{1}{2},\mathbf{j},2)}{3\Delta z^2}. \tag{11}$$

This kind of discretization provides a second order accuracy up to boundary cells. The convective terms do not need particular care near the boundaries; it is sufficient to substitute the assigned boundary value of the velocity components in (9).

Since this numerical method is partly implicit, boundary conditions are needed for $\hat{\mathbf{u}}$; in the original formulation [1] the pressure field was not introduced in the computation and from a Taylor expansion the boundary values were proved to be

$$\hat{\mathbf{u}}|_{\Gamma} = \mathbf{u}^{n+1} + \Delta t \mathbf{G}(p), \tag{12}$$

on the other hand pressure is explicitely introduced in the present formulation, thus

$$\hat{\mathbf{u}}|_{\Gamma} = \mathbf{u}^{n+1}. \tag{13}$$

48

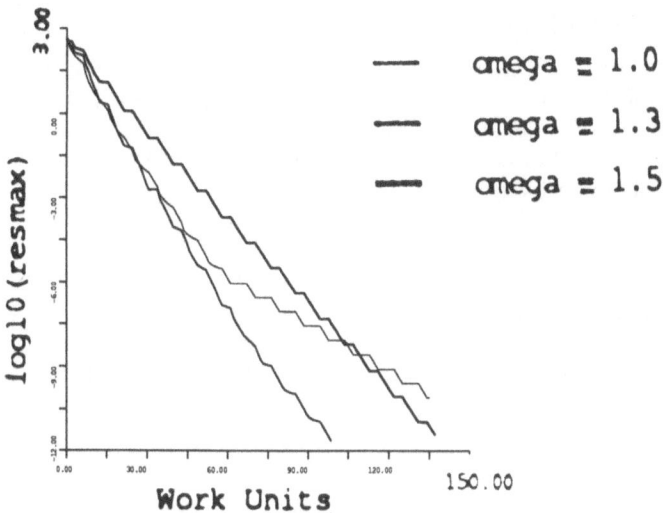

Figure 2: Convergence history of residual at the first time step.

3 Pressure solver

A realistic simulation of an unsteady incompressible flow requires the mass conservation equation to be verified accurately at each time step. Therefore a large fraction of total computing time must be devoted to this task, unless special solution techniques are introduced. Kim & Moin [1] used cosine FFT to solve the elliptic equation for Φ. In the present work a different approach has been chosen and the solution is obtained by means of a multigrid solver. The basic concept lying under a multigrid algorithm is that traditional iterative solvers are efficient in removing only high frequency error components; thus using different grid levels, it is possible to dump low frequency error components of the fine grid solution, which pretend to be high on coarser grids. Since equation (8) is linear a Correction Scheme Multigrid (CS) is used [6,7]. A multigrid algorithm is based on the following elements:

- an iterative solver used as smoothing operator;

- a restriction operator to transfer residual from fine to coarse grids;

- a prolongation operator to transfer corrections from coarse to fine grids.

3.1 Smoother

High frequency error components are dumped by a S.O.R. smoother. The value of the relaxation parameter ω is smaller than the optimum value for a single grid problem: $\omega = 1.3$ gives the best results for the $64 \times 64 \times 192$ grid used in the present simulation. In Figure 3 the convergence history of maximum residual of Equation (8) for different relaxation parameters is shown; the computational effort is expressed in work units (a work unit represents the cost of a single fine grid S.O.R. sweep). The convergence is brought up to 10^{-12} to show the decay rate of the residual; during the computation the limit has been increased to 10^{-4}.

3.2 Residual Restriction

The numerical scheme uses a cell-centered collocation for pressure; therefore fine and coarse grid have no common points (see Figure 3.2) and it is not possible to use direct injection for the evaluation of coarse grid residual; instead a full weighting procedure is used; if (ic, jc, kc) and (if = 2*ic-1, jf = 2*jc-1, kf = 2*kc-1) are coarse and fine grid points respectively

$$
\begin{aligned}
R(\text{ic}, \text{jc}, \text{kc}) = \\
[\quad R(\text{if}, \text{jf}, \text{kf}) \qquad &+\quad R(\text{if}+1, \text{jf}, \text{kf}) \qquad + \\
R(\text{if}, \text{jf}+1, \text{kf}) \quad &+\quad R(\text{if}+1, \text{jf}+1, \text{kf}) \qquad + \\
R(\text{if}, \text{jf}, \text{kf}+1) \quad &+\quad R(\text{if}+1, \text{jf}, \text{kf}+1) \qquad + \\
R(\text{if}, \text{jf}+1, \text{kf}+1) \quad &+\quad R(\text{if}+1, \text{jf}+1, \text{kf}+1) \quad]/8.
\end{aligned}
\tag{14}
$$

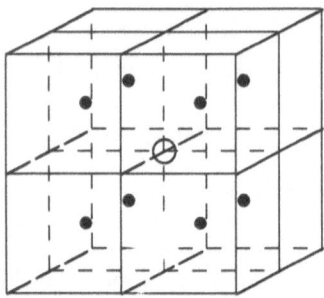

Figure 3: Fine • and coarse ∘ grid

3.3 Prolongation

The prolongation procedure is based on a bilinear interpolation scheme. If if = 2*ic, jf = 2*jc and kf = 2*kc are fine grid points, the coarse grid correction to the fine grid value $\Phi(\text{if}, \text{jf}, \text{kf})$ is given by

$$
\begin{aligned}
\Phi(\text{if}, \text{jf}, \text{kf}) = \Phi(\text{if}, \text{jf}, \text{kf}) + \\
[\quad 27\ \Phi(\text{ic}, \text{jc}, \text{kc}) \qquad &+\ 9\ \ \Phi(\text{ic}+1, \text{jc}, \text{kc}) \qquad + \\
9\ \ \Phi(\text{ic}, \text{jc}+1, \text{kc}) \quad &+\ 3\ \ \Phi(\text{ic}+1, \text{jc}+1, \text{kc}) \qquad + \\
9\ \ \Phi(\text{ic}, \text{jc}, \text{kc}+1) \quad &+\ 3\ \ \Phi(\text{ic}+1, \text{jc}, \text{kc}+1) \qquad + \\
3\ \ \Phi(\text{ic}, \text{jc}+1, \text{kc}+1) \quad &+\quad \Phi(\text{ic}+1, \text{jc}+1, \text{kc}+1) \quad]/64.
\end{aligned}
\tag{15}
$$

4 Discussion of results

The code ran on an IBM RS/6000-540 workstation equipped with 128 MB of RAM with the following parameters:

- grid size $N_x \times N_y \times N_z = 64 \times 64 \times 192$;

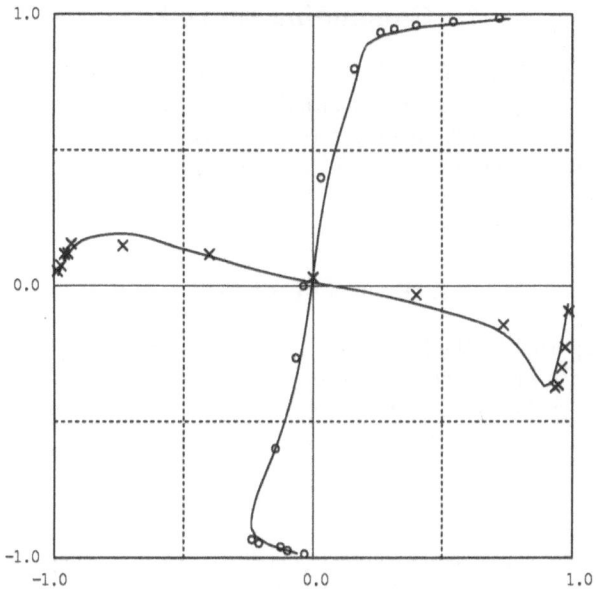

Figure 4: Experimental normalized velocity profiles u ($\circ$) and v ($\times$) in the symmetry plane $z = 0$, versus numerical profiles averaged over a period $\Delta T = 100$ (continuous line).

- time step $\Delta t = 0.0125$;

- Courant number $C \simeq 0.6$;

- total number of time steps NSTEP $= 16000$;

The computation has been performed with a residual of the Poisson equation fixed to 10^{-4} resulting in a divergence of the velocity field of the order of 10^{-8}. During initial stages the solution of Φ requires 4 multigrid cycles, but later on 3 cycles are enough to reduce the residual at the required level and the CPU time is 65 s per time step. During the simulation the flow maintains its symmetry across the plane $z = 0$, but since this fact was not supposed to be known apriori the whole cavity has been considered in the simulation.

In Figure 4 a comparison between numerical and experimental data (taken from Reference [2]) is shown; the agreement is quite good, but a longer averaging time could improve the numerical data.

The TGL vortices are shown in Figures 5 and 8; the velocity vector plots clearly show the mechanism of formation of these secondary instabilities: near the downstream separation eddy the primary circulation region behaves like a curved boundary layer, and the perturbation induced by the end-walls gives rise to an instability resulting in the TGL structures [8]. The number of vortex pairs changes during the evolution of the flow; at $T = 50$ nine pairs are clearly shown in the vorticity contour plots, while at $T = 200$ the vortex pair in the symmetry plane has disappeared. As TGL vortices move towards the symmetry plane, the flow in these regions is disturbed. The particles traces at $T = 50$ and $T = 200$ (see Figures 6 and 9) are completely different from the corresponding plots for the two-dimensional flow, which tends toward a steady state.

5   Concluding remarks

The numerical scheme seems to be effective in reproducing the physical behaviour of the flow. Furthermore the application of a multigrid method for the solution of the Poisson equation results in a very high speedup ratio. This fact makes complex computations feasible on workstations, without using any machine-dependent programming technique to avoid cache misses.

Acknowledgements

This work was supported by the Italian Ministry of Merchant Marine in the frame of the I.N.S.E.A.N. research plane 1990. The author would like to express gratitude to Paolo Orlandi of the University of Rome "La Sapienza" for useful discussions.

References

[1] Kim J., Moin P. "Application of a Fractional–Step Method to Incompressible Navier–Stokes Equations" *J. Comput. Phys.* **59** (1985), 309.

[2] Koseff J. R. "Momentum Transfer in a Complex Recirculating Flow" Stanford University, 1983

[3] Koseff J. R., Street R. L. *J. Fluids Eng.* **106** (1984), 21.

[4] Arakawa A. "Computational Design for Long-Term Numerical Integration of the Equations of Fluid Motion: Two-Dimensional Incompressible Flow." *J. Comp. Phys.* **1** (1966), 119.

[5] Perng C.-Y., Street R. L. "Three-Dimensional Unsteady Flow Simulations: Alternative Strategies for a Volume-Averaged Calculation" *Int. J. Numer. Methods Fluids* **9** (1989), 341.

[6] Brandt A. "Multi-level Adaptive Solutions to Boundary-Value Problems" *Math. Comput.* **31** (1977), 333.

[7] Hackbush W. "Multi-Grid Methods and Applications" Springer-Verlag, 1985.

[8] Drazin P.G., Read W.H. "Hydrodynamic Stability" Cambridge University Press, 1981.

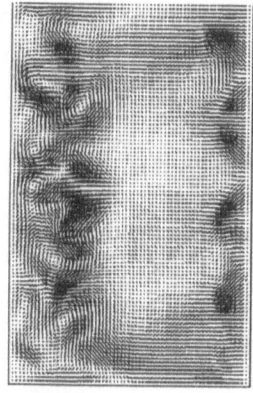 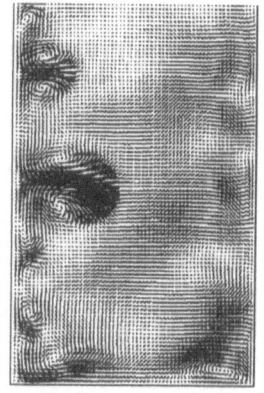

Figure 5: Velocity vector plot in the plane $y = 0$ at $T = 200$

Figure 8: Velocity vector plot in the plane $x = 0$ at $T = 200$

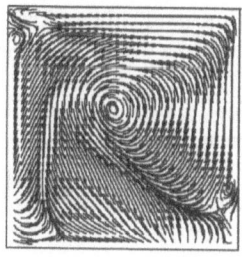

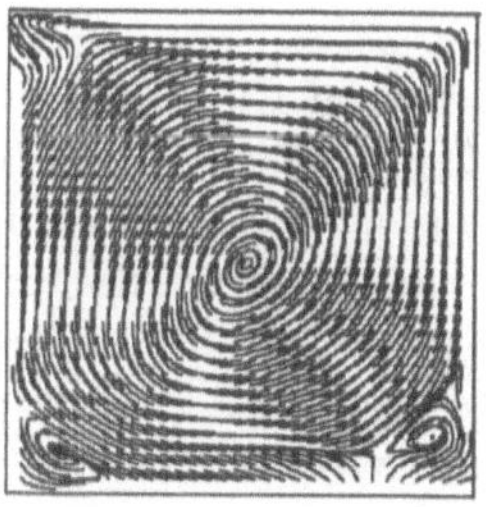

Figure 6: Particle traces in the plane $z = 0$ at $T = 50$

Figure 9: Particle traces in the plane $z = 0$ at $T = 200$

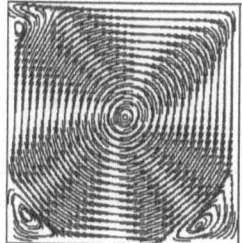 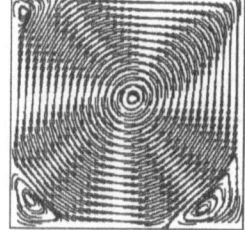

Figure 7: Particle traces at $T = 50$ in a two-dimensional cavity

Figure 10: Particle traces at $T = 200$ in a two-dimensional cavity

VELOCITY-VORTICITY SIMULATION OF UNSTEADY 3-D VISCOUS FLOW WITHIN A DRIVEN CAVITY

Y. Huang, U. Ghia

Department of Mechanical, Industrial and Nuclear Engineering

G. A. Osswald, K. N. Ghia

Department of Aerospace Engineering and Engineering Mechanics

Computational Fluid Dynamics Research Laboratory

University of Cincinnati, Cincinnati, Ohio 45221, USA

SUMMARY

A velocity-vorticity formulation of the unsteady three dimensional Navier-Stokes equations has been used to solve for the incompressible viscous flow within a driven cavity of spanwise aspect ratio 3:1 at a Reynolds number $Re = 3200$ on a non-uniform ($65 \times 65 \times 49$) grid covering one half of the span. An efficient Alternating-Direction-Implicit method which treats all cross derivative terms implicitly and which requires only six scalar tridiagonal sweeps has been developed for the vorticity transport equation. The divergence-curl formulation for the elliptic velocity problem is solved at each time step by a Multi-Grid Distributive Gauss-Seidel iterative scheme. The pressure is solved, only when desired, from the three-dimensional Pressure Poisson problem using a Multi-Grid Gauss-Seidel iterative scheme. Velocity, vorticity and pressure results are given for a characteristic time $t = 50$ after the upper surface of the cavity is impulsively started from rest.

INTRODUCTION

The velocity-vorticity $(\vec{V}, \vec{\omega})$ formulation of the unsteady three-dimensional Navier-Stokes equations has been selected for use in the present study, as opposed to the primitive variable velocity-pressure $(\vec{V}, p)$ formulation, for several specific reasons, foremost of which is the fact that the vorticity transport equation is quasi-linear in vorticity and independent of pressure, whereas the velocity transport equation (momentum equation) is nonlinear in velocity and coupled to the pressure. As transport is the critical physical phenomenon of unsteady viscous flows, the velocity-vorticity formulation yields a mathematically simpler, and more natural description in which the spin dynamics of a fluid particle (represented by the vorticity transport equation) is only loosely coupled to its kinematics (represented by the elliptic velocity problem) and is completely independent of the pressure obtained by solving the pressure Poisson problem.

FORMULATION

Let $\vec{V}(x, y, z, t)$ and $p(x, y, z, t)$ be the non-dimensionalized velocity vector and static pressure, respectively, for incompressible viscous flows in a three-dimensional domain $\mathcal{D}$ with $\partial\mathcal{D}$ as its boundary. The dimensionless unsteady incompressible Navier-Stokes equations in velocity-vorticity form are formulated as follows, neglecting body forces,

Vorticity Transport Equation

$$\frac{\partial\vec{\omega}}{\partial t} + \nabla \times (\vec{\omega} \times \vec{V}) + \frac{1}{Re}\nabla \times (\nabla \times \vec{\omega}) = 0 \qquad in \ \mathcal{D}. \tag{1}$$

Elliptic Velocity Problem

$$\nabla \bullet \vec{V} = 0 \qquad in \ \mathcal{D}, \tag{2}$$

$$\nabla \times \vec{V} = \vec{\omega} \qquad in \ \mathcal{D}, \tag{3}$$

subject to the compatibility constraint relationships that

$$\iint_{\partial\mathcal{D}} \vec{V} \bullet \vec{n} ds = 0 \tag{4}$$

and

$$\nabla \bullet \vec{\omega} \equiv 0 \qquad in \ \mathcal{D}. \tag{5}$$

Pressure Poisson Problem

$$\nabla^2 P = -\nabla \bullet (\frac{\partial\vec{V}}{\partial t} + \vec{\omega} \times \vec{V} + \frac{1}{Re}\nabla \times \vec{\omega}) \qquad in \ \mathcal{D} \tag{6}$$

where $P = p + \frac{1}{2}(\vec{V} \bullet \vec{V})$ is the total pressure.

All equations above are in strong conservation form; this property is maintained for the algebraic system of discretized equations as well.

SPACE-TIME DISCRETIZATION

A second-order accurate space-time discretization is employed, using central differences both in space and time, on the staggered grid arrangement shown in Fig. 1. Covariant velocity components are evaluated at the centroid of the cell surfaces in the computational domain, while covariant vorticity components are evaluated at cell edge midpoints. The pressure is evaluated at the cell centroid at the mid-time step. Evalution of the governing equations is staggered as well, with the ω_x transport equation being written at ω_x locations, the $\nabla \bullet \vec{V}$ equation being evaluated at the cell centroid and Eq. (3) being evaluated at the appropriate vorticity component locations. The pressure problem, Eq. (6), is evaluated at the cell centroid. Osswald et al. [1] have shown that this staggered discretization uniquely leads to a nonsingular matrix-vector formulation for the overdetermined elliptic velocity problem.

BOUNDARY CONDITIONS

Only the no-penetration condition enters the boundary conditions for the elliptic velocity problem, Eqs. (2-5), since only the normal velocity components appear at boundaries in the present staggered discretization. The no-slip condition affects vorticity creation at body surfaces and enters the vorticity transport analysis in the following manner. At body surfaces, vorticity is created, advected and diffused. The advection/diffusion of body vorticity is modelled by second-order accurate time extrapolation of corrected values of boundary vorticity. Using this time accurate estimate of future wall-vorticity and the vorticity transport equation (1), a time-accurate interior vorticity solution can be obtained. With interior vorticity known, the elliptic velocity problem, Eqs. (2-5), yields the instantaneous velocity field. Owing to the natural separation which occurs between the spin dynamics and the translational kinematics of the fluid particle, this interior velocity is independent of the current values of wall vorticity. Hence the no-slip condition can now be employed to create new wall vorticity, without altering the current velocity solution, and the calculations can proceed to the subsequent time level.

The boundary conditions for pressure are of the Neumann type; specifically,

$$\nabla P \bullet \vec{n} = -(\frac{\partial \vec{V}}{\partial t} + \vec{\omega} \times \vec{V} + \frac{1}{Re}\nabla \times \vec{\omega}) \bullet \vec{n} \qquad on \ \ \partial \mathcal{D} \tag{7}$$

where $\vec{n}$ is the outward unit vector normal to the boundary surface. The above boundary condition for pressure is such that the following integral constraint associated with the Poisson equation (6) is satisfied:

$$\iint_{\partial \mathcal{D}} \nabla P \bullet \vec{n} ds = \iiint_{\mathcal{D}} S_P dv, \tag{8}$$

where S_P represents the source term, namely, the right-hand side of Eq. (6) for the pressure Poisson problem. It is notable that, by using the pressure Poisson equation formulation, the pressure can be solved on any subdomain of the flow field or can be elected to be solved only at specific instances during the flow solution. Indeed, it need not be solved at all to compute the actual incompressible flow evolution. For incompressible flow fields, pressure only reacts to the vorticity transport and does not cause it.

SOLUTION METHODOLOGY AND STABILITY

The three components of the vorticity transport equation (1) are solved by a modified Douglas-Gunn variance of Alternating-Direction-Implicit (ADI) scheme developed by Osswald et al. [1, 2]. Implicit treatment of the cross-derivative vorticity terms is achieved within a six-sweep scalar tridiagonal inversion procedure which vectorizes completely. Because of the quasi-linear nature of the vorticity transport problem, a time step restriction can occur even with an implicit technique although at a level greater than the CFL restriction of explicit convective schemes. Round-off error growth can be optionally monitored within the present vorticity transport algorithm.

The elliptic velocity problem is solved by a Multi-Grid Distributive Gauss-Seidel (MG-DGS) iterative scheme developed by Huang et al. [3, 4]. It relaxes the overdetermined divergence-curl operator in Eq. (2, 3) for the instantaneous velocity field subject to the constraint equations (4, 5). A convergence history for the very first time step following the impulsive start from rest (a poor initial guess case) on the present ($65 \times 65 \times 49$) clustered grid is shown in Fig. 2. The convergence history of a single grid Distributive Gauss-Seidel (SG-DGS) technique for this same start-up case is also shown in Fig. 2 for comparison.

The pressure Poisson problem, Eq. (6), is solved by a Multi-Grid Gauss-Seidel (MG-GS) iterative procedure. Its convergence history is shown in Fig. 2 for the present ($65 \times 65 \times 49$) clustered grid. The pressure problem need be solved only when pressure information is desired and does not need to be solved to advance the flow evolution. The convergence history shown in Fig. 2 for the pressure is for the time step at $t = 42$ when the pressure solution is started.

TIME STEP AND SPACE GRID

A clustered ($65 \times 65 \times 49$) grid is employed within one half of the driven cavity with a symmetry condition being imposed along the spanwise midplane $z = 0$. Points are clustered near the walls with the greatest clustering near the top and bottom walls ($y = \pm\frac{1}{2}$) and near the downstream and upstream walls ($x = \pm\frac{1}{2}$). The maximal spacing in the $x - y$ planes is $(\Delta x)_{max} = (\Delta y)_{max} = 0.02323$; the minimal spacing is $(\Delta x)_{min} = (\Delta y)_{min} = 0.00781$. In the spanwise direction (z-direction) the maximal and minimal spacings are $(\Delta z)_{max} = 0.04916$ and $(\Delta z)_{min} = 0.01418$, respectively.

A constant time step of $\Delta t = 0.005$ was employed for the present solution.

COMPUTING COSTS

The solution was run on the CRAY Y-MP 8/864 at the Ohio Supercomputing Center using enhanced vectorization but *no* parallel implementation at present. The ($65 \times 65 \times 49$) clustered grid distribution yields $1,210,368$ degrees of freedom when no pressure solution is computed and an additional $196,608$ degrees of freedom for a full pressure field, totalling to $1,406,976$ degrees of freedom to solve for $(\vec{V}, \vec{\omega}, p)$.

With a time step of $\Delta t = 0.005$, 10^4 time steps were required to reach a characteristic time $t = 50$. This required a total of 6.88 CPU hours, with the pressure solution computed only between $t = 42$ and $t = 50$. The averaged computational efficiency was $1 \times 10^{-5} sec/gridpoint/timestep$ to determine $(\vec{V}, \vec{\omega})$ for $t = 0 - 42$, and 4×10^{-5} to obtain $(\vec{V}, \vec{\omega}, p)$ for $t = 42 - 50$.

RESULTS

The detailed simulation results are presented here for the shear-driven cavity flow configuration with aspect ratio 3 : 1 and Re=3200, using a nonuniform ($65 \times 65 \times 49$) grid at characteristic time $t = 50$. The streamwise velocity profile $u(y)$ along the line $x = 0, z = 0$ is depicted in Fig. 3, which also shows evidence of three-deminsionality in the behavior of this profile in the proximity of the bottom surface $y = -\frac{1}{2}$. The corresponding vertical, i.e., transverse, velocity profile $v(x)$ along the line $y = 0, z = 0$ is presented in Fig. 4, where strong downflow near the front surface $x = \frac{1}{2}$ is clearly seen. The profiles of pressure $p(x)$ and $\omega_z(x)$ along the centerline $z = 0$ of the top surface $y = \frac{1}{2}$ are shown in Fig. 5 and Fig. 6, whereas the pressure $p(z)$ and vorticity $\omega_x(z)$ along the line given by $x = 0$ and $y = 0$ are delineated in Figs. 7 and 8, respectively.

Figure 9 presents the time history of the pressure $p(t)$ at the midpoint of the top surface $x = 0, y = \frac{1}{2}, z = 0$ for time between $t = 42$ and $t = 50$; this shows significant unsteadiness. On the other hand, the time history for the vorticity ω_z at the same location is plotted for the entire time, i.e., between $t = 0$ and $t = 50$ in Fig. 10. The apparent gap in this time history between $t = 22$ and $t = 26$ is due to loss of a data set which was accidently erased, leaving this gap. The tangential, i.e., streamwise, velocity components for the spanwise midplane $z = 0$ are plotted in Fig. 11; the three-dimensional effect in the flow is, again, very vividly seen. The contours of the vorticity component ω_z and static pressure are shown in Figs. 12 and 13, respectively. The development of strong shear layers is very evident from Fig. 12.

At the longitudinal plane $x = 0.26508$ near the front wall of the driven cavity, the tangential velocity vectors at various spanwise locations are depicted in Fig. 14. In this same plane, the contours of vorticity ω_x and static pressure p are presented in Figs. 15 and 16. The $7\frac{1}{2}$ pairs of counter-rotating Taylor-Görtler vortices which exist across the half span are seen in Fig. 15. In this as well as in the figures to follow, due to the assumed symmetry about the midplane $z = 0$, the simulation results are displayed for only half the span. i.e., for $-\frac{3}{2} \leq z \leq 0$. For the cavity midplane $x = 0$, the tangential velocity is depicted in Fig. 17, whereas the contours of vorticity are plotted in Fig. 18; these show $6\frac{1}{2}$ pairs of Taylor-Görtler vortices across the half-span. the pair nearest the endwall in Fig. 15 has diffused. Also, each pair has been displaced vertically upward from the bottom surface with an oppositely rotating pair of vortices induced locally beneath each Taylor-Görtler pair. The static pressure contours are shown in Fig. 19. The third vertical plane considered is at $x = -0.47115$ near the back wall of the driven cavity; here again, the same three quantities are plotted. The tangential velocity is shown in Fig. 20. The contours of vorticity ω_x, given in Fig. 21, show little structure in them. It should be noted that, at this rear wall location, the vorticity ω_y is expected to have most of the spanswise structure.

Results are also given for the mid-plane $y = 0$ in Figs. 23-25. The tangential velocity vectors are illustrated in Fig. 23 at various spanwise locations. The contours of vorticity ω_y are plotted in Fig. 24 and show a complex eddy pattern that exists near the back wall, $x = -\frac{1}{2}$, while the Taylor-Görtler vortices appear to be just developing along the front wall, $x = \frac{1}{2}$, at this instant. The contours of static pressure are given in Fig. 25. Finally, the contour of vorticity ω_z and pressure are also demonstrated at the top surface $y = \frac{1}{2}$ in Figs. 26 and 27, respectively.

CONCLUSION

An efficient numerical algorithm for unsteady three-dimensional incompressible Navier Stokes equations in the form of vorticity-velocity variables has been developed and presented. A Multi-Grid method has been developed to solve the pressure Poisson problem with Neumann boundary conditions. The method has proved to be efficient and robust. The core memory requirement is about 7 megawords for the $(65 \times 65 \times 49)$ grid. The methodology is applied to the simulation of unsteady 3-D incompressible flow within the shear-driven cavity. The flow at Reynolds number $Re = 3200$ is persistently unsteady through at least $t = 50$. The Taylor-Görtler instability is shown to play a key role in generating three-dimensionality in the evolving flow.

References

[1] Osswald, G. A., Ghia, K. N. and Ghia, U., *AIAA No. 87-1139, AIAA 8th Comp. Fluid Dynamics Conf.,* Honolulu, Hawaii, June 1987.

[2] Osswald, G. A., Ghia, K. N. and Ghia, U., *Proc. 11th Int. Conf. Num. Methods Fluid Dynamics,* Williamsburg, Virginia, July 1988.

[3] Huang, Y. and Ghia, U., *Proc. Fifth Copper Mtn. Conf. Multigrid Techniques,* Copper Mountain, Colorado, April 1991.

[4] Huang, Y., Ghia, U., Osswald, G. A. and Ghia, K. N., *AIAA No. 91-1562, Proc. AIAA 10th Comp. Fluid Dynamics Conf.,* Honolulu, Hawaii, June 1991.

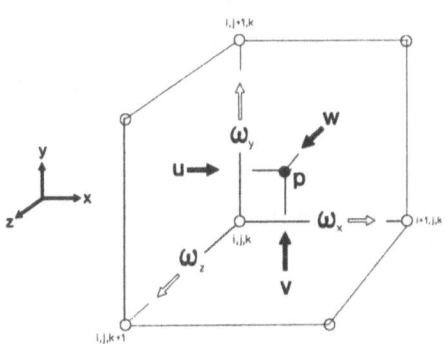

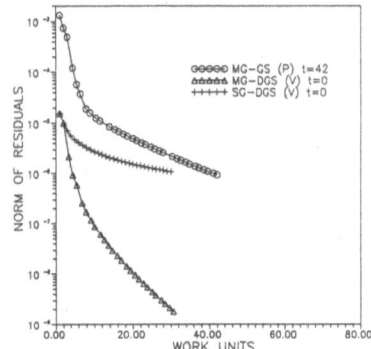

FIG. 1. COMPUTATIONAL CELL AND STAGGERED GRID FOR VELOCITY, VORTICITY AND PRESSURE.

FIG. 2. CONVERGENCE HISTORY FOR VELOCITY AND PRESSURE PROBLEM FOR (65x65x49) CAVITY AT Re = 3,200.

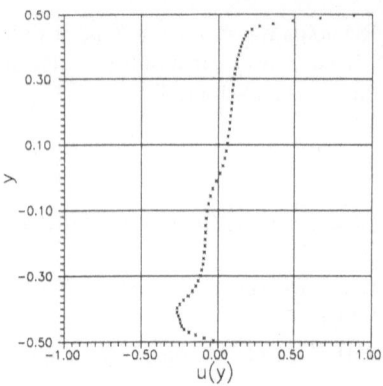

FIG. 3. HORIZONTAL VELOCITY PRO-
FILE u(y) AT z=0, x=0 and t=50.
Re = 3,200, (65x65x49) GRID.

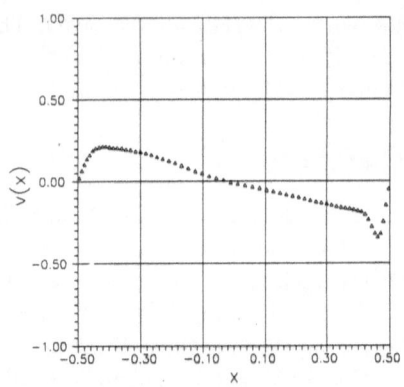

FIG. 4. VERTICAL VELOCITY PROFILE
v(x) AT z=0, y=0 and t=50.
Re = 3,200, (65x65x49) GRID.

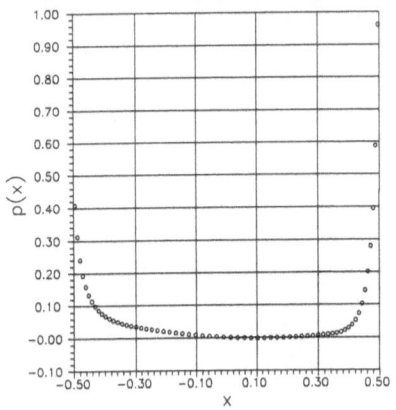

FIG. 5. PRESSURE PROFILE p(x) AT
z=0, y=1/2 and t=50. Re = 3,200,
(65x65x49) GRID.

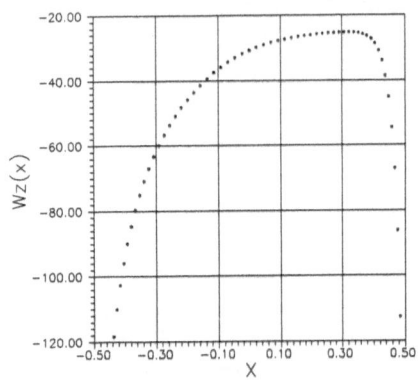

FIG. 6. PROFILE OF VORTICITY COM-
PONENT ω_z(x) AT z=0, y=1/2 and t=50.
Re = 3,200, (65x65x49) GRID.

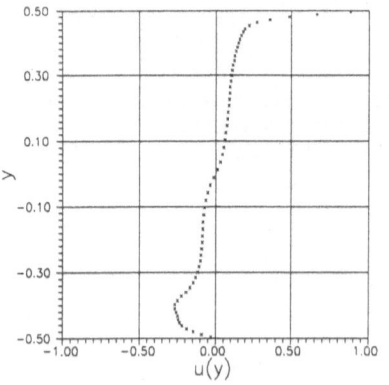

FIG. 7. PRESSURE PROFILE p(z) AT
x=0, y=0 and t=50. Re = 3,200,
(65x65x49) GRID.

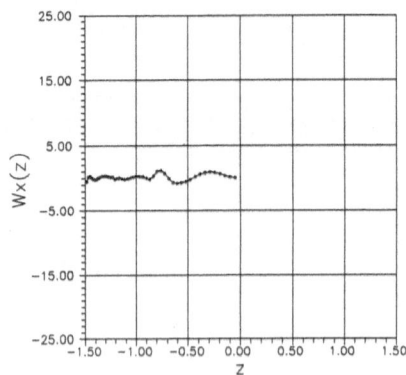

FIG. 8. PROFILE OF VORTICITY COM-
PONENT ω_x(z) AT x=0, y=0 and t=50.
Re = 3,200, (65x65x49) GRID.

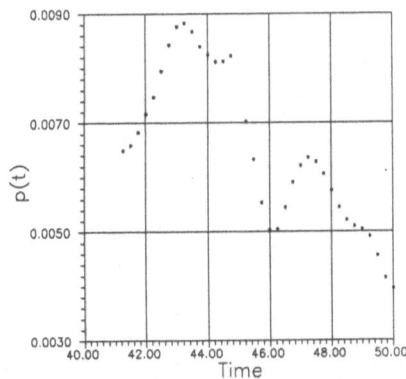

FIG. 9. TIME HISTORY OF PRESSURE
p(t) AT x=0, y=1/2 and z=0 FOR
t=42-50. Re = 3,200, (65x65x49)
GRID.

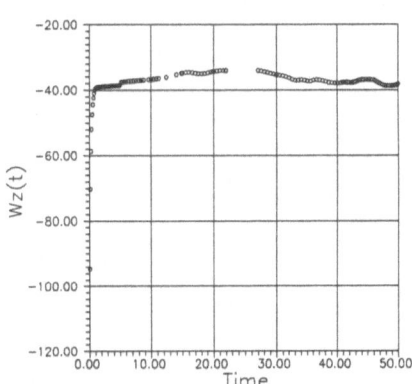

FIG. 10. TIME HISTORY OF VORTICITY
COMPONENT $\omega_z(t)$ AT x=0, y=1/2, z=0
FOR t=0-50. Re = 3,200,(65x65x49)
GRID.

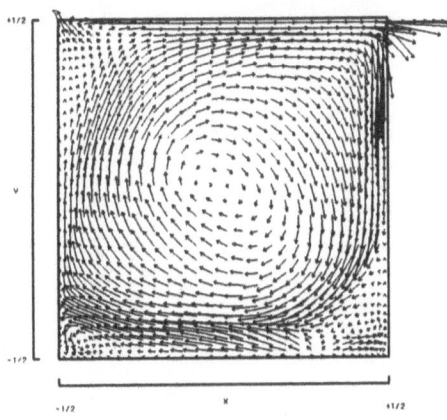

FIG. 11. VERTICAL/HORIZONTAL VAR-
IATION OF TANGENTIAL VELOCITY COM-
PONENT AT MIDPLANE z=0.

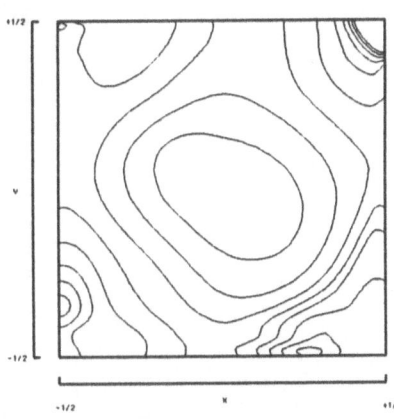

FIG. 13. VERTICAL/HORIZONTAL VARI-
ATION OF PRESSURE AT MIDPLANE z=0.
ϕ_0=-0.0271, ϕ_1=0.0299.

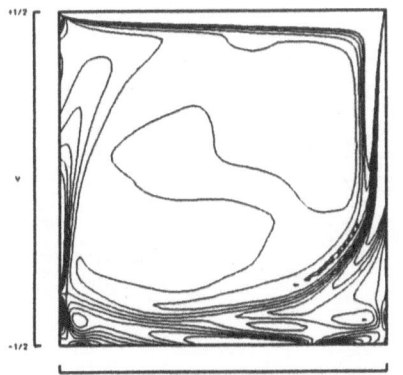

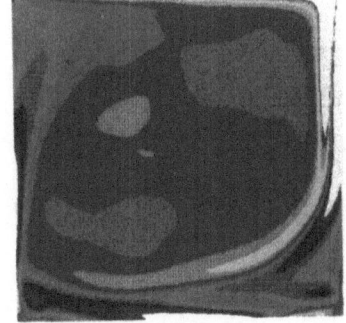

FIG. 12. VERTICAL/HORIZONTAL VARIATION OF NORMAL VORTICITY COMPONENT
ω_z AT MIDPLANE z=0. ϕ_0=-5, ϕ_1=5.

61

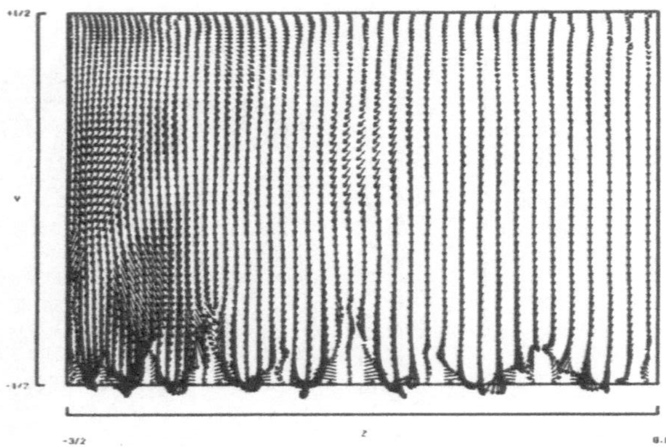

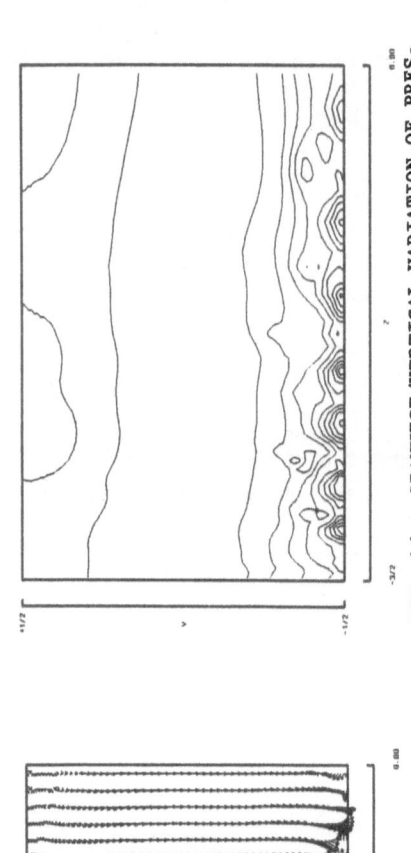

FIG. 16. SPANWISE/VERTICAL VARIATION OF PRESSURE NEAR FRONT WALL =0.26508. ϕ_0=-0.0181, ϕ_1=0.0608.

FIG. 14. SPANWISE/VERTICAL VARIATION OF TANGENTIAL VELOCITY COMPONENTS NEAR FRONT WALL x=0.26508.

FIG. 15. SPANWISE/VERTICAL VARIATION OF NORMAL VORTICITY COMPONENT ω_x NEAR FRONT WALL x=0.26508. ϕ_0=-10, ϕ_1=10.

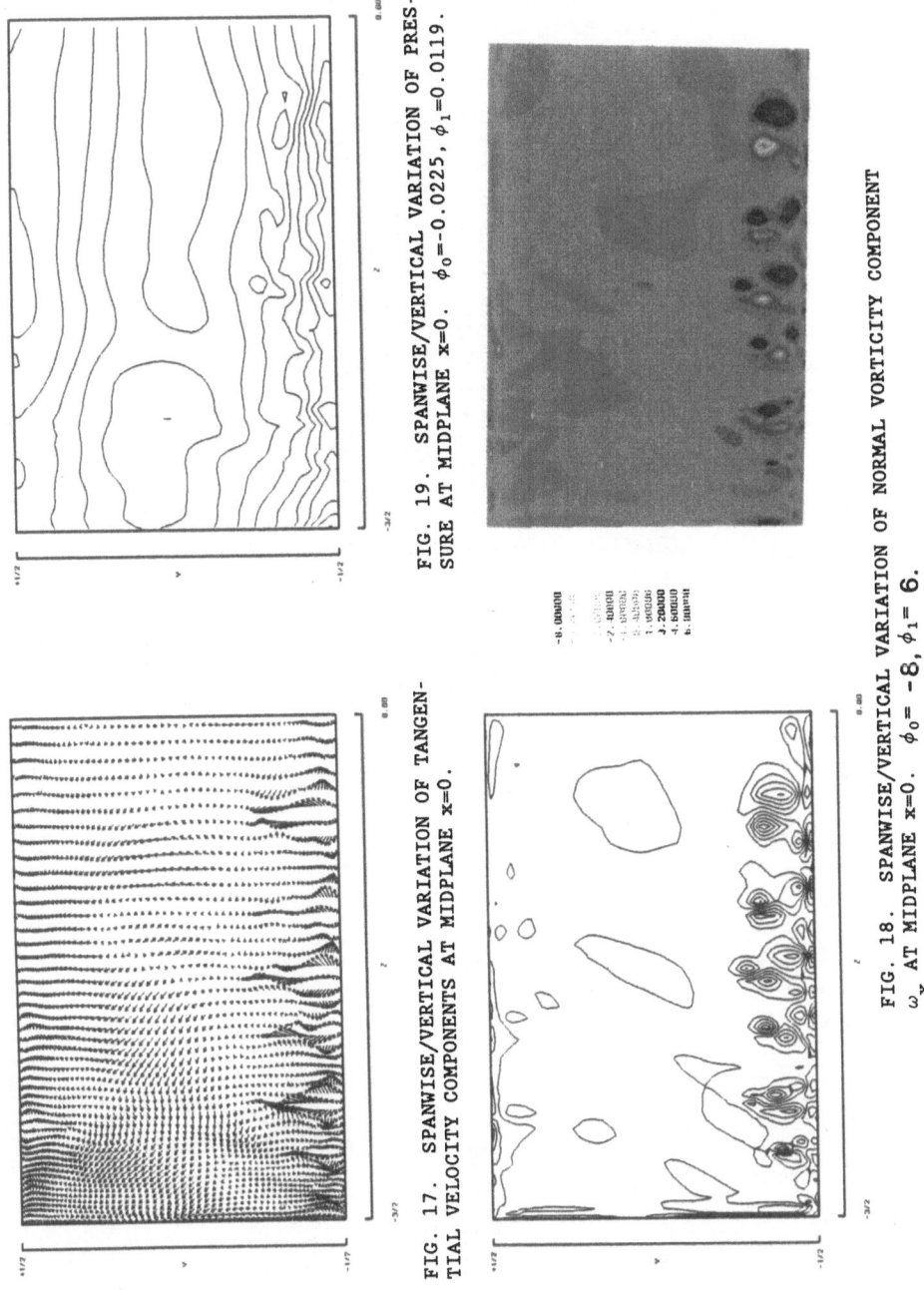

FIG. 17. SPANWISE/VERTICAL VARIATION OF TANGENTIAL VELOCITY COMPONENTS AT MIDPLANE x=0.

FIG. 18. SPANWISE/VERTICAL VARIATION OF NORMAL VORTICITY COMPONENT ω_x AT MIDPLANE x=0. $\phi_0=-8$, $\phi_1=6$.

FIG. 19. SPANWISE/VERTICAL VARIATION OF PRESSURE AT MIDPLANE x=0. $\phi_0=-0.0225$, $\phi_1=0.0119$.

63

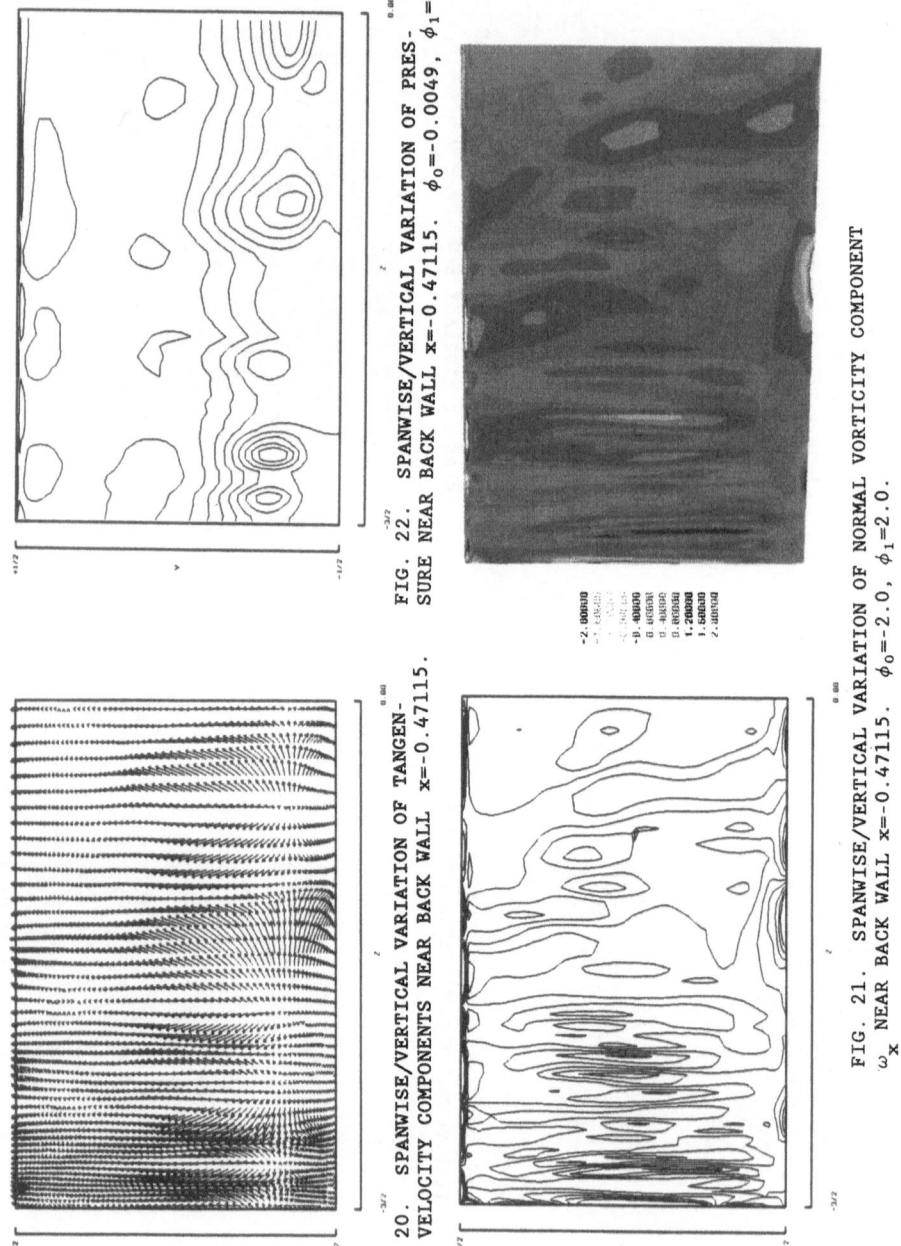

FIG. 22. SPANWISE/VERTICAL VARIATION OF PRES-
SURE NEAR BACK WALL x=-0.47115. $\phi_0=-0.0049$, $\phi_1=0.0292$.

FIG. 20. SPANWISE/VERTICAL VARIATION OF TANGEN-
TIAL VELOCITY COMPONENTS NEAR BACK WALL x=-0.47115.

FIG. 21. SPANWISE/VERTICAL VARIATION OF NORMAL VORTICITY COMPONENT
ω_x NEAR BACK WALL x=-0.47115. $\phi_0=-2.0$, $\phi_1=2.0$.

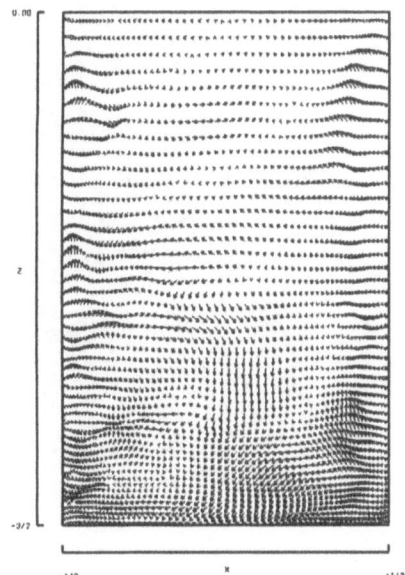

FIG. 23. SPANWISE/HORIZONTAL VARI-
ATION OF TANGENTIAL VELOCITY COMPO-
NENTS AT MIDPLANE y=0.

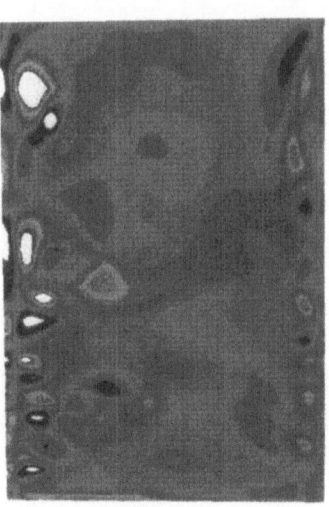

FIG. 25. SPANWISE/HORIZONTAL VARI-
ATION OF PRESSURE AT MIDPLANE y=0.
ϕ_0=-0.0224, ϕ_1=0.0014.

FIG. 24. SPANWISE/HORIZONTAL VARIATION OF NORMAL VORTICITY COMPONENT
ω_y AT MIDPLANE y=0. ϕ_0=-2.0, ϕ_1=2.0.

65

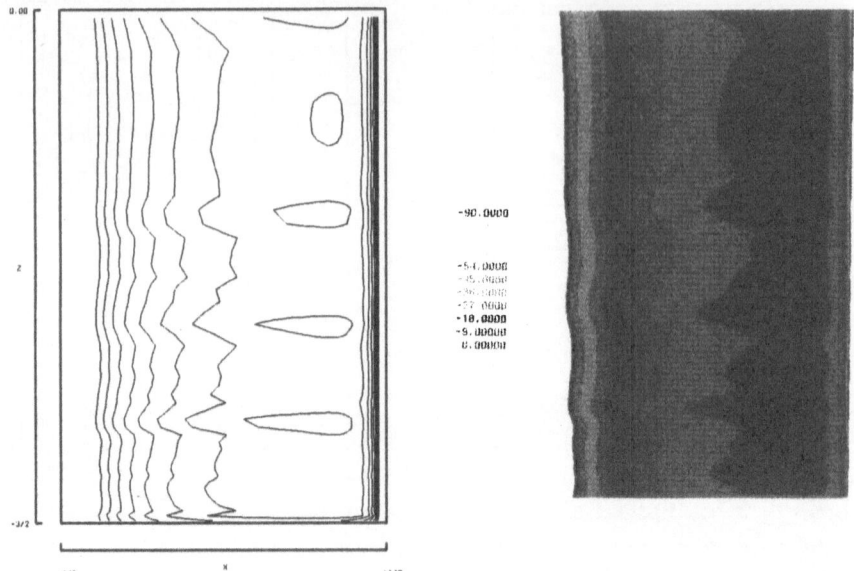

FIG. 26. SPANWISE/HORIZONTAL VARIATION OF VORTICITY COMPONENT ω_z AT TOP PLANE y=1/2. ϕ_0=-90, ϕ_1=0.0.

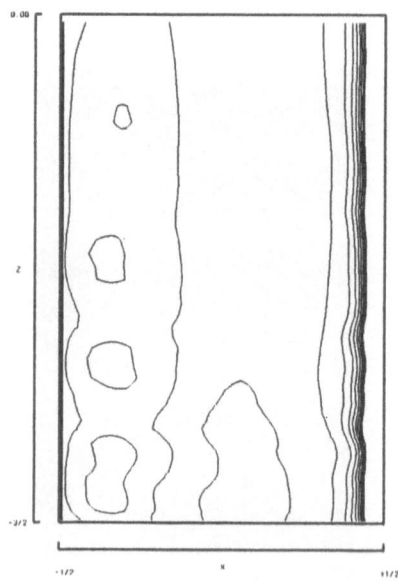

FIG. 27. SPANWISE/HORIZONTAL VARI-ATION OF PRESSURE AT TOP PLANE y=1/2. ϕ_0=-0.0293, ϕ_1=0.0776.

A 3-D DRIVEN CAVITY FLOW SIMULATION WITH N3S CODE

L. JANVIER*, B. METIVET**, R. MGOUNI*, G. POT* and E. RAZAFINDRAKOTO**

ELECTRICITE DE FRANCE

ABSTRACT

We present in this paper the calculation of a flow in a 3-D driven cavity, with the help of the finite element thermal-hydraulic code N3S, which has been developed at EDF. Computations have been realized in a half-cavity, without turbulence modelling. Two time schemes based on characteristics method have been used : the full computations (up to the time t = 200 s.) have been done with a first order time scheme, while a second order time scheme has been tested for a partial computation (up to t = 50 s.).

INTRODUCTION

At the present time, the finite element code N3S is solving the 2-D or 3-D Navier-Stokes equations, coupled to energy equation either via the Boussinesq's approximation. Furthermore, the k-epsilon model can be used in the turbulence cases. Indeed we have to compute industrial flows where the Reynolds number may reach several millions.

Though finite difference or finite volume codes based on structured meshes– like ESTET – are also developed at EDF, a finite element code can deal with very complex geometries and allows local mesh refinements. Moreover, the finite element method is well suited to theoretical analysis. However the matrix systems involved are more expensive to solve and the memory storage is bigger.

The driven cavity test-case must be better suited to a high order structured mesh code, because of the simplicity of the geometry. However it's of great interest to us to compute it with N3S to improve the validation of the code. Let us add that the number of degrees of freedom has been chosen as a compromise between the computation cost and the accuracy.

In Section I, we present briefly the numerical method used to solve the incompressible Navier-Stokes equations in the laminar case, with emphasis on the time scheme which has been recently improved.

In Section II, we give some details about the numerical implementation and the jagged-diagonal matrix storage.

The numerical results relative to the test-case are detailed in Section III.

* EDF, Département Laboratoire National d'Hydraulique, 6 Quai Watier, 78 400, Chatou, FRANCE
** EDF, Département Mécanique et Modèles Numériques, 1, Avenue du Général de Gaulle, 92 141, Clamart, FRANCE

I. DISCRETIZATION SCHEMES

The time scheme is based on an operator splitting method. The first step handles with the convection part with the help of the characteristics method, while the second step deals with the Stokes problem which is solved by a standard finite element Galerkin formulation. The finite elements which are available in the code are the P1-P2 and isoP1-P1 velocity-pressure triangle and tetrahedron that verify the inf-sup condition and also the Q1-Q2 quadrangle and brick ([1], [2]).

In order to improve the accuracy, we have recently implemented besides the original time scheme ([3], [4]) which is of order 1, a second order time scheme that has been proposed in [5], [6], [7]. One of its advantages is that it preserves a good stability behaviour.

We denote by $\Omega \subseteq \mathbf{R}^N$ (N = 2 or 3) the computational domain and by [0, T] the time interval. We consider the Navier-Stokes equations : find $u : \Omega \to \mathbf{R}^N$ and $p : \Omega \to \mathbf{R}$ solutions of

$$\frac{\partial u}{\partial t} + u \cdot \nabla u - \nu \Delta u + \nabla p = f \qquad \text{in } \Omega \times]0, T[, \tag{I.1}$$

$$\text{div } u = 0 \qquad \text{in } \Omega \times]0, T[, \tag{I.2}$$

$$u (x, 0) = u_0 (x) \qquad \text{for } x \in \Omega , \tag{I.3}$$

$$u = u_d \qquad \text{on } \Gamma \times]0, T[. \tag{I.4}$$

For the sake of simplicity, we only present the case of Dirichlet boundary conditions and assume that the boundary velocity u_d is tangent to Γ.

I.1 DESCRIPTION OF THE TIME SCHEMES

Let Δt be the time step and $t^{n+1} = (n+1)\Delta t$. We compute approximations $U^{n+1} : \Omega \to \mathbf{R}^N$ and $P^{n+1} : \Omega \to \mathbf{R}$ of the velocity and the pressure at time t^{n+1} in the following way.

Concerning the convection step, for any point $x \in \Omega$, we introduce the characteristic curve $\mathbf{X}_x^{n+1} : [t^{n*}, t^{n+1}] \to \mathbf{R}^N$, solution of

$$\begin{cases} \forall t \in [t^{n*}, t^{n+1}[, \quad \dfrac{d\mathbf{X}_x^{n+1}}{dt} (t) = U^{n*} (\mathbf{X}_x^{n+1} (t)), \text{ if } \mathbf{X}_x^{n+1} (t) \in \overline{\Omega}, \\ \qquad\qquad\qquad\qquad\qquad 0, \qquad\qquad \text{otherwise,} \\ \mathbf{X}_x^{n+1} (t^{n+1}) = x , \end{cases} \tag{I.5}$$

where

(i) for the first order scheme:
$t^{n*} = t^n$ and $U^{n*} = U^n$,

(ii) for the second order scheme:
$t^{n*} = t^{n-1}$ and $U^{n*} = 2 U^n - U^{n-1}$.

Then for any point $\mathbf{x} \in \Omega$, we set for both schemes:
$$\widetilde{U}_1^{n+1}(\mathbf{x}) = U^n(X_{\mathbf{x}}^{n+1}(t^n)),$$
(I.6)
and for the second order scheme only :
$$\widetilde{U}_2^{n+1}(\mathbf{x}) = U^{n-1}(X_{\mathbf{x}}^{n+1}(t^{n-1})).$$
(I.7)

Note that the second order scheme consists in computing a characteristic curve on an interval twice as long as the one computed with the first order one and in calculating two convected fields instead of one. Thus, this step is two times more expensive .

The Stokes step consists in computing (U^{n+1}, P^{n+1}) solution of

(i) for the first order scheme :

$$\left\{ \begin{array}{ll} \dfrac{U^{n+1} - \widetilde{U}_1^{n+1}}{\Delta t} - \nu\,\Delta\,U^{n+1} + \mathbf{grad}\ P^{n+1} = f\,(\,.\,,\,t^{n+1}) & \text{in } \Omega, \\[2ex] \text{div } U^{n+1} = 0 & \text{in } \Omega, \\[1ex] U^{n+1} = u_d\,(.,\,t^{n+1}) & \text{on } \Gamma. \end{array} \right.$$
(I.8)

(ii) for the second order scheme:

$$\left\{ \begin{array}{ll} \dfrac{\frac{3}{2}\,U^{n+1} - 2\,\widetilde{U}_1^{n+1} + \frac{1}{2}\,\widetilde{U}_2^{n+1}}{\Delta t} - \nu\,\Delta\,U^{n+1} + \mathbf{grad}\ P^{n+1} = f\,(\,.\,,\,t^{n+1}) & \text{in } \Omega, \\[2ex] \text{div } U^{n+1} = 0 & \text{in } \Omega, \\[1ex] U^{n+1} = u_d\,(.,\,t^{n+1}) & \text{on } \Gamma. \end{array} \right.$$
(I.9)

I.2 SPACE DISCRETIZATION

We consider the full space-time discretization and we denote by (U_h^{n+1}, P_h^{n+1}) the velocity-pressure approximation. The convected fields $\widetilde{U}_{i,h}^{n+1}$, $i = 1, 2$, are computed according to (I.6) and (I.7) for any velocity node **a**. The discrete convected fields are thus defined by interpolation :

$$\widetilde{U}_{i,h}^{n+1}(\mathbf{x}) = \sum_{\mathbf{a} \text{ velocity node}} \widetilde{U}_{i,h}^{n+1}(\mathbf{a})\ \varphi_{\mathbf{a}}(\mathbf{x}), \text{ for } i = 1, 2,$$

where $\varphi_{\mathbf{a}}$ is the shape function at the velocity node **a**.

I.3 SOME SCHEME PROPERTIES

In this section, we only consider problems (I.1)-(I.4), the solution (u, p) of which is sufficiently smooth. Moreover, we assume that $u_d = 0$.

I.3.a Time consistency error

The following properties are proved in [8]. It is easy to see that only the momentum conservation equation produces a non zero consistency error.

For any point $x \in \Omega$, let $\chi_x^{n+1} : [t^{n-1}, t^{n+1}] \to R^N$ denote the characteristic curve defined by a system analogous to (I.5), where the discrete velocity is replaced by u. We also set

$h_x(t) = v(\chi_x^{n+1}(t), t)$ for $t \in [t^{n-1}, t^{n+1}]$.

The consistency error is equal to ($\| . \|$ is the euclidian norm of R^N):

$$E(v, p) = \text{Sup} \{ \| e(x, n+1) \|, \ x \in \Omega, n / t^{n+1} \in [0, T] \},$$

where, for the first order scheme:

$$e(x, n+1) = -(\Delta t) [\frac{1}{2} \frac{d^2 h_x}{dt^2}(t^{n+1}) + \frac{\partial v}{\partial t} . \nabla v(x, t^{n+1})] + O([\Delta t]^2),$$

and for the second order scheme:

$$e(x, n+1) = -(\Delta t)^2 [\frac{1}{3} \frac{d^3 h_x}{dt^3}(t^{n+1}) + \frac{\partial^2 v}{\partial t^2} . \nabla v(x, t^{n+1})] + O([\Delta t]^3).$$

We have shown on computations [8] that the second order scheme improves the velocity and, above all, the pressure in the case of dominating advection flows. Moreover it allows a better simulation of the unsteady flows.

I.3.b Stability results

By using techniques similar to those of [5], we have proved [9] the following result concerning the **second order scheme** in the case of the P1-P2 or P1-isoP1 elements, when the convected fields are computed according to (I.6)-(I.7) without any interpolation error (see § I.2).

Proposition: *There exist constants* C_1, C_2, C_3 *and a constant* h_0 *sufficiently small, all being independent of* h *and* Δt, *such that the condition*

$h \leq h_0$ *and* $\Delta t \leq C_3 h^{1/2}$

yields to

$\| U_h^n \|_{0,\infty} \leq C_1$ *and* $\| U_h^n \|_2 \leq C_2, \ \forall n / n \Delta t \leq T.$

The previous result is concerned with unsteady flow computations — the steady case (when the solution is achieved for n going to infinity) is under consideration. Note that we get from [3] that the **first order scheme** is unconditionally stable when no space discretization is used. The case of the fully discretized scheme will soon be studied. From now on, experience shows a good stability behaviour for Courant numbers less than 10.

II. IMPLEMENTATION

The discretized Stokes problem is solved with the help of a preconditioned Uzawa algorithm (gradient method on the equivalent pressure system, described in details in [1]). For this laminar test-case, the chosen preconditioning is a combination of the inverses of the pressure Laplacian and Mass operators, which is optimal in a certain way ([10]).

Each of the elementary linear system (on velocity or pressure) is solved by a preconditioned conjugate gradient algorithm. The storage mode of the different matrices is a jagged diagonal one, which allows a speed-up of 12 in 2D and 6 in 3D on matrix-vector product in comparison with the classical symmetric compressed row storage ([11]).

III - NUMERICAL RESULTS

III.1 COMPUTATIONS

Computations performed with the first order scheme are referred to as case A, the ones performed with the second order scheme as case B.

III.1.a Geometry

The aim of the workshop is to compute a three-dimensional lid-driven cavity flow. This flow has been experimentally studied at Stanford (see [12]). The whole cavity has a width W of 1m (x-axis), a depth D of 1m (y-axis) and a lateral span L of 3m (z-axis). On account of the observations and computations reported in [12], we have assumed that the flow was symmetric about the centre plane of the lateral span (z =0). The cavity has been divided into two parts and the computation realized in the right hand side (figure III - 1) The computational domain extends from x = - 0.5 (upstream) to x = 0.5 (downstream), from y = - 0.5 (bottom) to y = 0.5 (top) and from z = 0 (symmetry) to z = 1.5 (wall). For this flow, the Reynolds number based on D is 3200.

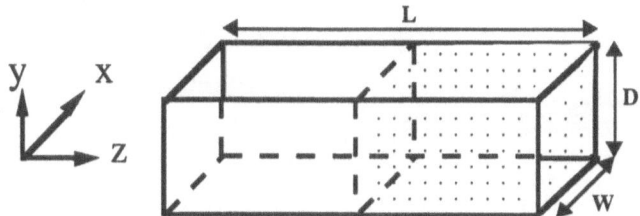

Figure III-1 : Schematic view of the computational domain area

III.1.b Boundary conditions

At the top, the velocity $U = (u, v, w)$ is given : $U = (1, 0, 0)$. On the wall boundaries, no slip conditions are imposed : $U = (0, 0, 0)$. On the symmetry plane, the normal component

w is zero and the conditions $\dfrac{\partial u}{\partial z} = \dfrac{\partial v}{\partial z} = 0$ are imposed implicitly thanks to the variational

Galerkin formulation.

III.1.c Space Grid

A nonuniform mesh has been carried out with P_1-P_2 tetrahedra (figures III-2 and III-3). For the half-cavity, it contains 36330 elements, 52661 velocity nodes (corresponding to 41, 41, 31, in each space direction), and 7166 pressure nodes.

This mesh has been realized with a grid point distribution built from the one suggested for an ERCOFTAC Benchmark for turbulent natural convection in a square cavity. It provides a refinement near the walls with a not too sharp progression from one element to its neighbours (1. to 1.62). If we denote by (x_i, y_j, z_k) $1 \leq i,\ j \leq imax$, $1 \leq k \leq kmax$ the coordinates of the pressure nodes, we have :

$$\frac{x_i}{W} = \frac{y_i}{D} = \frac{1}{2}\left[1 + \frac{\tanh\ [\alpha_1(\,i\,/\,imax\ -1/2)]}{\tanh\ (\alpha_1\,/\,2\,)} \right] - \frac{1}{2}, \qquad i = 0,\ 1,...imax,$$

$$\frac{z_k}{L} = \frac{1}{2}\left[\frac{\tanh\ (\alpha_2\,k\,/\,kmax)}{\tanh\ (\alpha_2)} \right], \qquad k = 0,\ 1,...kmax,$$

where $\alpha_1 = 5.$; $\alpha_2 = 1.55$; $imax = 20$; $kmax = 15$.

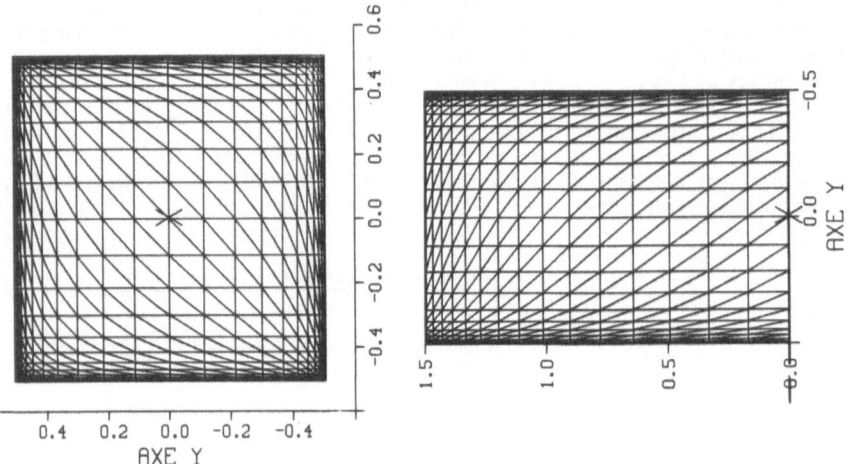

Figure III-2 : plane XY   Figure III-3 : plane YZ

III.1.d Time step

The computational time step has been chosen equal to 0.02 s, which corresponds to a maximum Courant number of 3.0 (in fact, case A had been started with a 0.01s time step (up to t=1s), it had been set to 0.05 to accelerate the run (up to t = 4s), then reduced to 0.02).

72

III.1.e Computing costs

The simulation has been performed on a CRAY Y-MP 4128. It has needed less then 18 megawords of memory. At each time step, the average number of iterations for the Uzawa algorithm, equal to 6, has led to a less than 10^{-7} velocity field divergence (see § II).

For the case A, the last required output time ($t_3 = 200$) has been obtained after more than forty hours of simulation. The average CPU time is 16,50 s. per time step, 10^{-4} s. per degree of freedom and time step. Concerning the case B, the first output time ($t_1 = 50$), has been obtained after nearly 20 hours (corresponding to 28.16 s. per time step and 1.7 10^{-4} s. per degree of freedom and time step). Thus our implementation of the second order time scheme increased the time cost by 70%, but the matrix preconditionings are not yet optimized in this last case. As these costs include the initialization stages for each restart run, they may be overestimated.

III.2 ANALYSIS OF THE RESULTS

The computed results are given at the times $t_1 = 50$, $t_2 = 100$, $t_3 = 200$ for the case A and at the time $t_1 = 50$ for the case B.

III.2.a Velocity profiles in the symmetry plane

Figures III-4A and III-4B show u versus y at line x=0 and v versus x at line y=0. They can be compared with the experimental profiles obtained in [13, mentionned in 12].

Concerning the case A, a comparison between the three profiles shows that the velocity evolves with respect to the time. The third output is the closest to the [13] data. At first sight, the case B shows a quicker transient evolution : indeed the first time is closer to the case A third one. Moreover, apart from the bottom of the cavity, where the computed boundary layer is too thick, results B/t_1 and A/t_3 are in good agreement with the experimental results.
It must be pointed out that the experimental results are time averaged ones, so, as far as no agreement is known whether the flow reaches a steady state or not, we can't validate the comparisons.

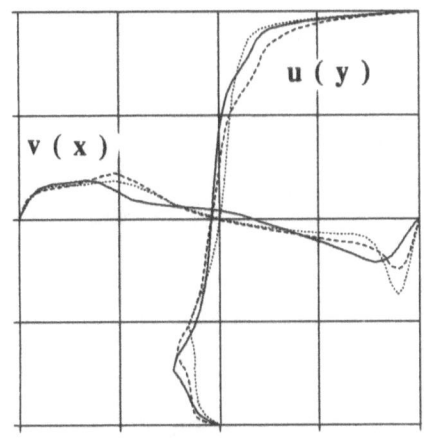

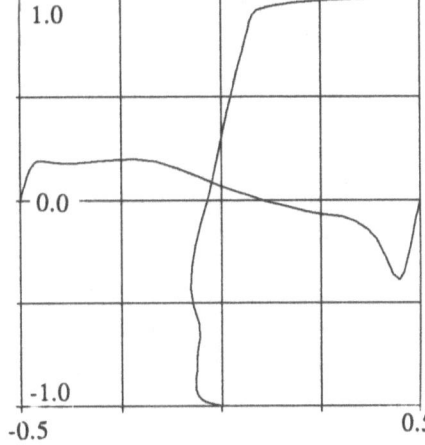

FIGURE III-4A   FIGURE III-4B

— : t=50 ; ---- : t = 100 ; ···· : t = 200

III.2.b   x - vorticity profiles

Figures III-5A and III-5B show x-vorticity profiles versus z along the line x=0, y=0 . We would have expected to count the vortices developed in the z-direction. We can't count a constant number of zero-line crossing. In order to investigate the transient stage, we have added on the figure III-5A, the same profile a the times of 180 and 190. We can see that change of sign does occur, particularly near plane z=0. (see below, III.2.f).

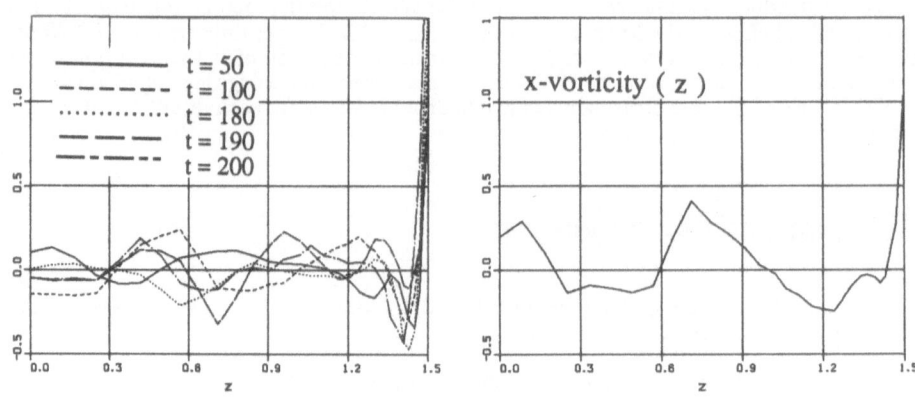

FIGURE III-5B

III.2.c Time evolutions

Figures III-6 shows the evolution in time of pressure and of the z-component of the vorticity ω at the point x=0, y=0.5, z=0 located at the middle of the top line of the symmetry plane. We notice that the stage is always transient at time t = 200.

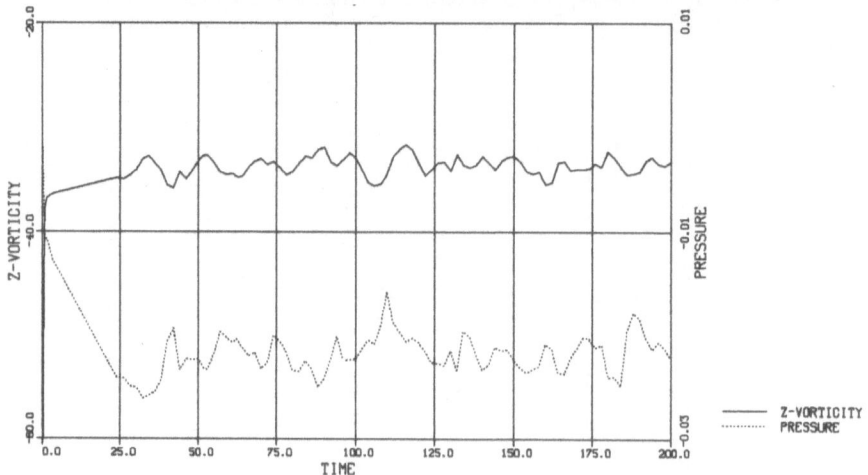

FIGURE III-6

III.2.d Contours of vorticity in the symmetry plane

Figures III-7A and III-7B represent contours of z component of the vorticity ω, at the plane $z = 0$. A main eddy is obtained and secondary ones in the corners, more or less developed, depending on the time of plotting. The location of the center of the main eddy must be unsteady (see below).

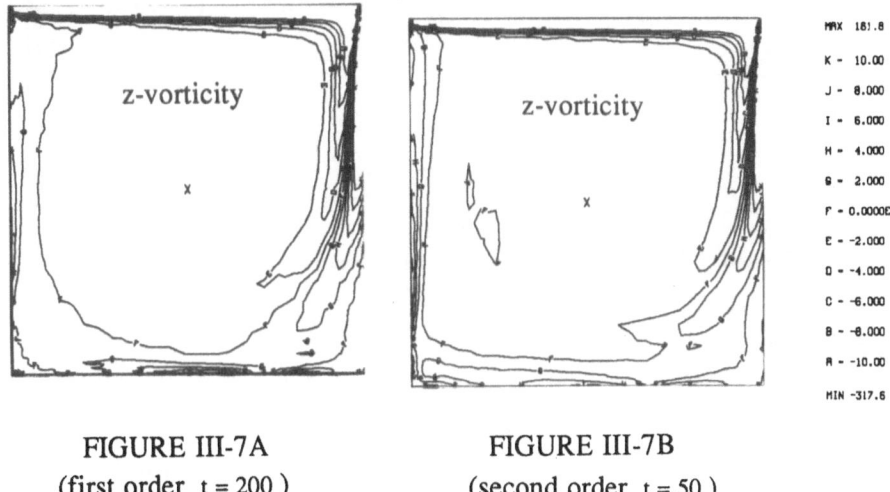

MAX 181.8

K - 10.00
J - 8.000
I - 6.000
H - 4.000
G - 2.000
F - 0.0000E
E - -2.000
D - -4.000
C - -6.000
B - -8.000
A - -10.00

MIN -317.6

| FIGURE III-7A | FIGURE III-7B |
| (first order $t = 200$) | (second order $t = 50$) |

III.2.e Contours of vorticity in x = cst planes

Figure III-8 represents, for the first order scheme at time t_1, contours of x-vorticity on the downstream plane $x = 4/15$, the middle plane $x = 0$, and the upstream plane $x = -7/15$. The results are given at time t_3 for the first order scheme. Vertical eddies can be seen (as expected in III.2.b), downstream, where z-axis vortices are spread, they are centered closer to the bottom than upstream. A similar behaviour is observed in the second order computations.

III.2.f Velocity fields

The velocity fields, drawn on current lines, are shown in the symmetry plane $z = 0$ (figure III-9) and in the upstream plane $x = -7/15$ in order to see how vortices are organized, and how they are varying with respect to the time. The [12] authors suggest that the location and the size of the vertical vortices are "dynamic and possibly periodic". The structure of our computed flow may confirm that, at least we notice a significant interaction between z-vortices (the "main" ones) and x-vortices (the vertical, or Taylor-Görtler-like, ones).

CONCLUSION

To conclude, we must first precize that, because of the choice of a half cavity as a computational domain, the results can be somewhat different from those obtained in the entire cavity (see for example the number of vortices on figure III-9). Furthermore, to do this computation, we have used N3S code in a standard way of utilization, without, for example, increasing too much the total number of discretization nodes.

At least, concerning our results, we can notice that the flow (computed with first order time scheme) looks still unsteady even at time t=200s (figure III-6).

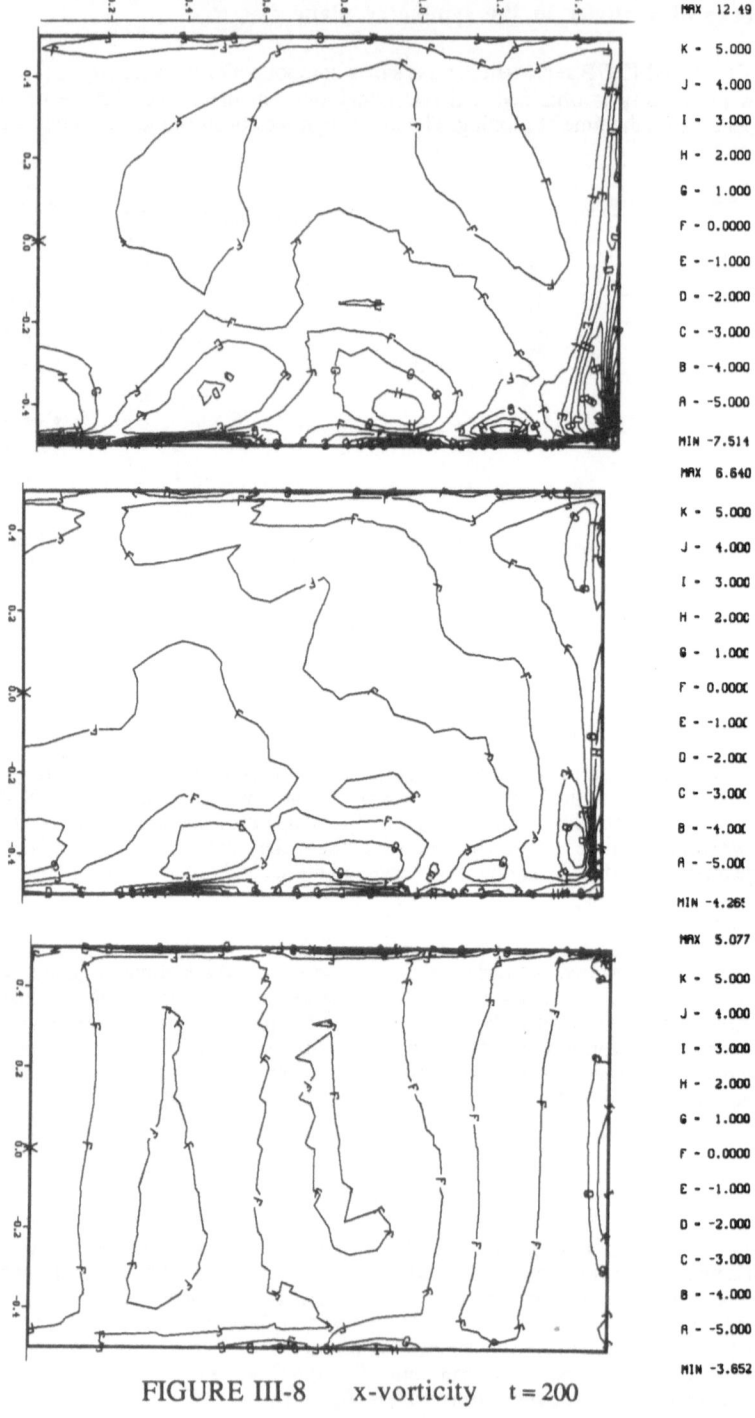

MAX 12.49

K - 5.000
J - 4.000
I - 3.000
H - 2.000
G - 1.000
F - 0.0000
E - -1.000
D - -2.000
C - -3.000
B - -4.000
A - -5.000

MIN -7.514

MAX 6.640

K - 5.000
J - 4.000
I - 3.000
H - 2.000
G - 1.000
F - 0.0000
E - -1.000
D - -2.000
C - -3.000
B - -4.000
A - -5.000

MIN -4.265

MAX 5.077

K - 5.000
J - 4.000
I - 3.000
H - 2.000
G - 1.000
F - 0.0000
E - -1.000
D - -2.000
C - -3.000
B - -4.000
A - -5.000

MIN -3.652

downstream plane x = 4/15

middle plane x = 0

upstream plane x = -7/15

FIGURE III-8   x-vorticity   t = 200

76

symmetry plane z=0   upstream plane x=-7/15

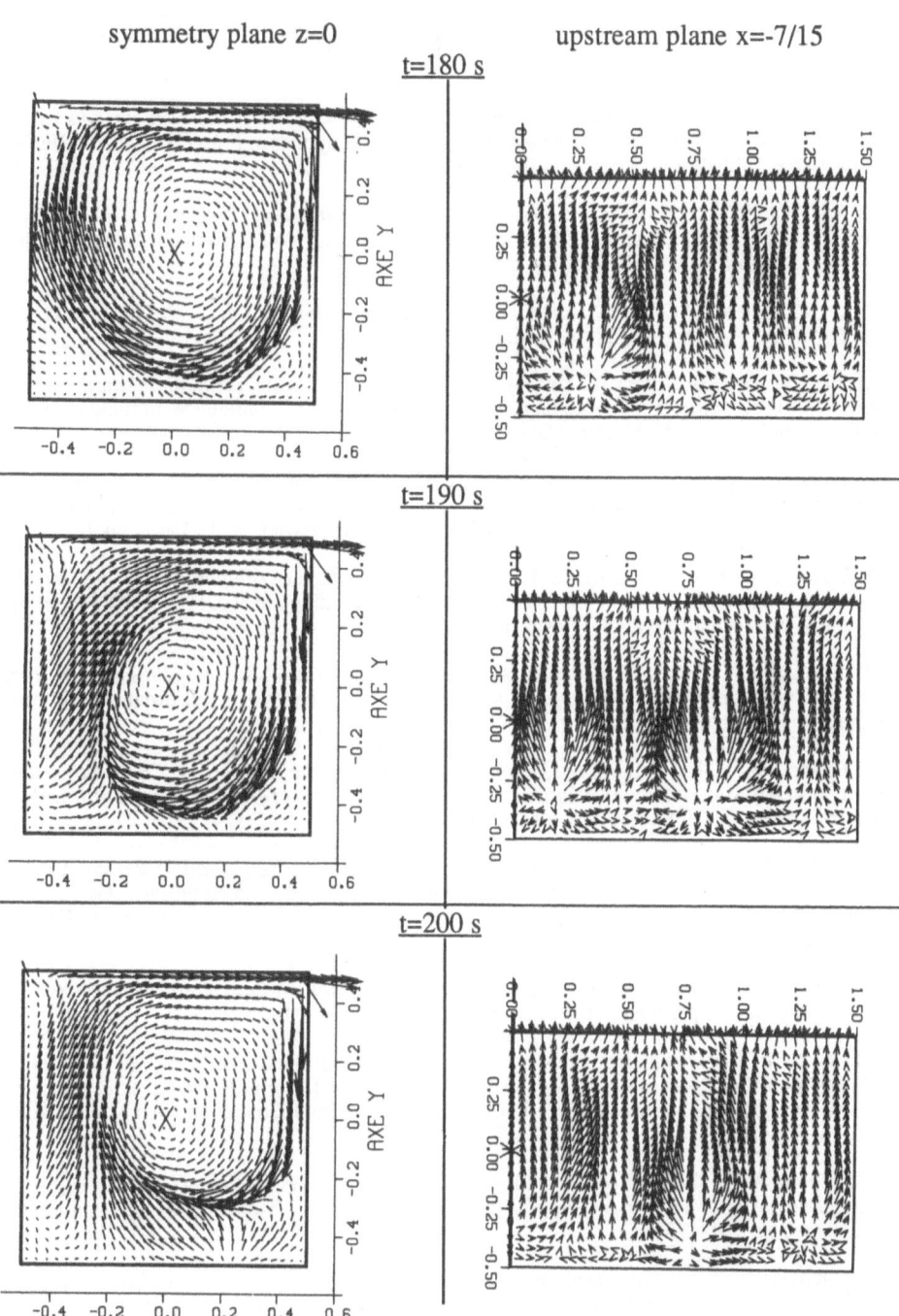

Figure III.9

REFERENCES

[1] J.P. Chabard,B. Métivet, G. Pot, B. Thomas, *"An Efficient Finite Element Method for the Computation of 3D Turbulent Compressible Flows"* , to appear in Finite Element in Fluids. Hemisphere Publishing Corporation Ed.

[2] J.P. Chabard, *"A finite element method for turbulent incompressible flows"* Computational Fluid Dynamics for Industrial Flows Series, Von Karman Institute for Fluid Dynamics Lectures, April 1990.

[3] O. Pironneau, *"On the transport diffusion algorithm and its applications to the Navier-Stokes equations."* , Numer. Math. 38, 309-332, 1982.

[4] J.P. Benque, B. Ibler, A. Keramsi, G. Labadie, *"A finite element method for the Navier-Stokes equations."*, Proceedings of the third international conference on finite elements in flow problems. Banff.Alberta, Canada, 10-13 June 1980.

[5] R. E. Ewing, T. F. Russel, *"Multistep Galerkin Methods along Characteristics for Convection-Diffusion Problems."* , Advances. in Comp. Meth. for P.D.E., R. Vichnevetsky & R.S. Stepleman eds., IMACS, Rutgers Univ., New Brunswick, N.J., 1981, pp 28-36.

[6] L Ho, Y. Maday, A. Patera, E. Ronquist, *"A high order Lagrangien decoupling method for the incompressible Navier-Stokes equations."* Proceedings of ICOSAHOM'89 meeting, C.Canuto & A. Quarteroni eds, North Holland (1990).

[7] Y. Maday, A. Patera, E. Ronquist, *"An operator integration factor splitting method for time dependant problems. Application to incompressible fluid flows."* Journal of Scientific Computing, December 1990.

[8] B. Métivet, E. Razafindrakoto,*"Projet N3S de mécanique des fluides. Etude numérique d'un schéma aux caractéristiques d'ordre 2 pour la résolution des équations de Navier-Stokes."* EDF Report ref. HI72/7094 (1990).

[9] K. Boukir, Y. Maday, B. Métivet, E. Razafindrakoto, *"Presentation of a second order scheme using the characteristics method applied to the Navier-Stokes equations."* Proceedings of the 13th IMACS World Congress on Computation and Applied Mathematics, Dublin, July-91.

[10] J. Cahouet, J.-P. Chabard, *" Some fast three dimensionnal finite element solver for the generalized Stokes problem "* Int. J. Num. Math. Fluids, Vol. 8, PP. 869-895, 1988.

[11] B. Gest, J. P. Grégoire, B. Nitrosso, G. Pot, *" Conjugate gradient's performances enhanced in the C.F.D. code N3S by using a better vectorization of matrix vector product"* Proceedings of the 13th IMACS World Congress on Computation and Applied Mathematics, Dublin, July 91.

[12] Freitas C. J., Street R. L., Findikakis A. N., Koseff J. R. : *"Numerical Simulation of Three-Dimensional Flow in a Cavity"*, International Journal for Numerical Methods in Fluids, vol 5, (1985), 561-575.

[13] Koseff J. R. : *"Momemtum transfer in a complex recirculating flow"*, Ph D Dissertation, Departement of Civil Engineering, Stanford University, California, 1983.

NUMERICAL SIMULATION OF THREE-DIMENSIONAL UNSTEADY FLOW IN A CAVITY

A. Kost, N.K. Mitra, M. Fiebig
Institut für Thermo- und Fluiddynamik
Ruhr-Universität Bochum
Universitätsstr. 150, D-4630 Bochum, FRG

SUMMARY

The three-dimensional unsteady flow of an incompressible viscous fluid in a rectangular cavity is considered. The flow is driven by the sliding upper wall of the cavity. A finite-volume method with non-staggered variable arrangement is used to solve the three-dimensional, time-dependent Navier-Stokes equations. Due to the influence of the end-walls the flow structure exhibits substantial differences from two-dimensional solutions. The method employs higher-order discretization for the convection fluxes in order to resolve the experimentally observed unsteady features of the flow. Results for a cavity flow at a Reynolds number of 3200 show good agreement with previously reported numerical and experimental results.

INTRODUCTION

Recirculating flows frequently occur in engineering applications. The lid-driven flow in a rectangular cavity can serve as an ideal representation of such flows because of its simple geometry and the straightforward boundary conditions. In recent years, systematic studies by both experiments and numerical methods have been carried out [1-7]. Numerical simulation of the experimentally observed Taylor-Görtler-like (TGL) longitudinal vortices has been shown to depend strongly on the accuracy and efficiency of the numerical procedure. Higher-order accurate discretization of the convective terms and sufficient grid resolution are necessary to simulate successfully the significant three-dimensional features of the laminar lid-driven cavity flow [4-7].

In the present paper, the flow of an incompressible viscous fluid in a rectangular lid-driven cavity is considered. Fig. 1 shows the geometry and flow definitions. The flow is driven by the moving upper wall of the cavity which produces a Reynolds number of 3200 (based on the speed of the upper wall and the cavity width). Initially the flow is at rest. No-slip and impermeable boundary conditions apply to the solid walls. The cavity has a depth-to-width aspect ratio of D/B=1 and a lateral span-to-width aspect ratio of L/B=3.

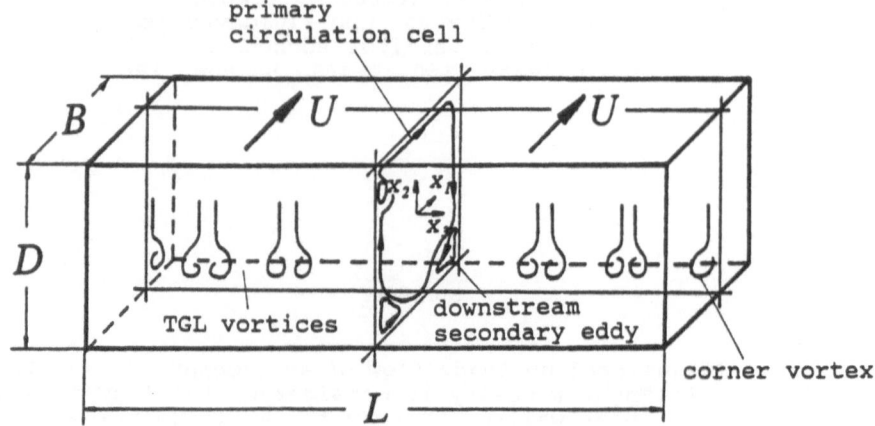

Fig. 1: Geometry and qualitative character of the lid-driven cavity flow

NUMERICAL PROCEDURE

Governing equations and basics of the code

The equations governing the flow of an incompressible viscous fluid are the well-known Navier-Stokes equations, which may be written in conservative form:

$$\frac{\partial(\rho\Phi)}{\partial t} + \frac{\partial}{\partial x_j}(C_j\Phi - \Gamma\frac{\partial\Phi}{\partial x_j}) = S_\Phi \qquad j=1,2,3 \qquad (1)$$

where $C_j = \rho u_j$ and $\Phi=1$, $\Gamma=0$ for the continuity equation and $\Phi=u_i$, $\Gamma=\mu$ for the momentum equation. S_Φ contains the appropriate source terms. p is the pressure, ρ is the density and μ is the dynamic viscosity of the fluid.

Most flows in engineering practice are three-dimensional and have rather complex boundaries. The general requirements of an advanced numerical procedure to be used in practical calculations should therefore be that it can be applied to general flow situations and that it uses an efficient solution algorithm with good convergence characteristics. Recently, finite-volume methods for solving flow equations on curvilinear boundary fitted grids employing non-staggered variable arrangement and Cartesian velocity components have been proposed [8,9]. For the present method these concepts have been adopted.

80

Grid generation

The geometry of a rectangular cavity permits us to use a simple orthogonal grid. One-dimensional grid clustering is employed in order to resolve adequately the velocity gradients in the boundary layers. For instance, we have in the x_1 direction [6]:

$$x_1 = \frac{0.5 \tanh(a\xi)}{\tanh(a)} ; \qquad \xi = -1 + \frac{2(i-1)}{i_{max}-1} ; \qquad 1 \le i \le i_{max} \qquad (2)$$

$$a = 0.5 \log(\frac{1+b}{1-b}) ; \qquad b = 0.9 . \qquad (3)$$

Similar transformations are used in the x_2 and x_3 direction. For the present study we use a 25x25x75 grid, resulting in a minimum spacing between adjacent grid nodes of 0.014 near the solid walls and a maximum spacing of 0.065 in the cavity center.

Discretization procedure

Integration of the momentum equations over a finite number of control volumes (CV) leads to the balance equations of momentum fluxes I through the CV faces and volumetric sources S, thus (cf. Fig. 2 for grid arrangement and nomenclature):

$$I_e - I_w + I_n - I_s + I_t - I_b = S . \qquad (4)$$

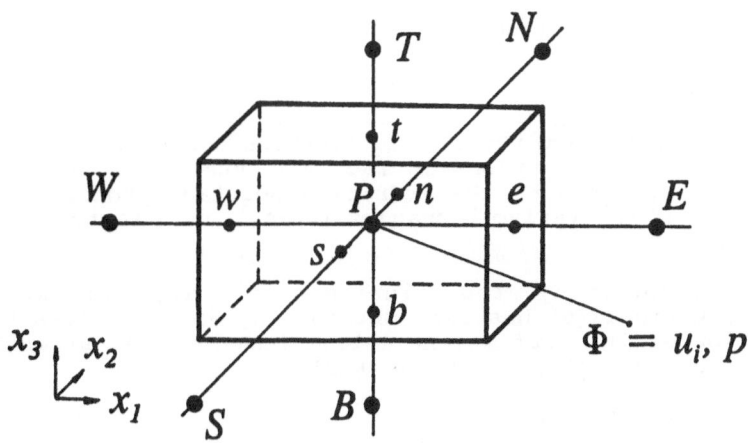

Fig. 2: Grid arrangement and nomenclature

The total flux through each CV face consists of two distinct contributions: the convection flux I_C and the diffusion flux I_D. The diffusion flux can simply be obtained by assuming linear variation of the variables between adjacent grid points. Evaluation of the convective part requires discretization schemes for interpolating variables at the cell face from their nodal values.

In order to reduce the spurious numerical diffusion introduced by a first-order upwind differencing scheme a number of higher-order upwind schemes have been developed [8]. The quadratic upwind differencing scheme (QUDS) uses an upstream-weighted quadratic interpolation for the convection fluxes. It was used in recent numerical simulations of the three-dimensional unsteady flow in lid-driven cavities [4-7]. In the present code the QUDS has been implemented by treating explicitly all contributions from nodes other than the essential ones (i.e. P,E,W,N,S,T,B, cf. Fig. 2). This procedure not only reduces the order of the coefficient matrix, but also enhances the stability of the iterative solution method [8].

Experimental flow visualizations [1-3] showed that the structure of the TGL vortices varies strongly in time. The discretization of the time-dependent term of eq. (1) has therefore been chosen to be second-order accurate:

$$[\frac{\partial(\rho\Phi)}{\partial t}]^{n+1} = \frac{\rho}{2\Delta t}(3\Phi^{n+1}-4\Phi^n+\Phi^{n-1}),$$ (5)

Replacing the expressions for the convective and diffusive parts of the cell face fluxes and the corresponding source terms, the discretized transport equation for a CV centered about node P takes the following linearized form:

$$\frac{a_P}{\alpha_\Phi}\Phi_P = \sum a_{nb}\Phi_{nb}+b_\Phi+\frac{1-\alpha_\Phi}{\alpha_\Phi}a_P\Phi_P^* \qquad nb=E,W,N,S,T,B$$ (6)

where the coefficients a_{nb} represent the combined convection and diffusion effects and the term b_Φ contains the discretized source terms. Since the equations in fact are non-linear and coupled, for the convergence of the iterative solution procedure under-relaxation of the variable changes ($\Phi_P-\Phi_P^*$) by a factor $0<\alpha_\Phi\leq 1$ is employed.

In the present study, the solution of the algebraic system of linear equations (6) has been obtained by an incomplete LU factorization of the coefficient matrix based on the strongly implicit procedure (SIP) of Stone [10].

Pressure-velocity coupling

For incompressible flows, the convergence of the numerical method for solving the momentum and continuity equation depends strongly on an adequate handling of the pressure-velocity coupling. The velocity field obtained by solving the momentum equations using a guessed pressure field in general does not satisfy the continuity equation. The continuity equation does not involve the pressure but serves as an additional constraint for the velocity field. In the present code the coupling between pressure and velocities is achieved by the well-known SIMPLEC algorithm [11]. In order to avoid an oscillatory pressure field resulting from non-staggered arrangement of the variables a special interpolation practice has been used to determine the mass fluxes F through CV faces. As proposed by Perić [8] the discretized momentum equations serve as the basis for this interpolation. For evaluation of the velocities at CV faces all terms except the pressure difference across the face are interpolated linearly from the nodal values on either side of the face, e.g. for the "e" face:

$$u_{1e}^* = \overline{\left(\frac{\sum a_{nb} u_{nb}^* + b_u^*}{a_P}\right)_e} - \overline{\left(\frac{1}{a_P}\right)_e} A_e (p_E^* - p_P^*). \tag{7}$$

Here, the superscript asterisk denotes values obtained by solution of the momentum equations with a guessed pressure field p^*, the overbar indicates linear interpolation, A_e is the cell face area and b_u^* is the rest of the corresponding source term excluding the pressure term, which has been written explicitly. The velocity components u_{2n}^* and u_{3t}^* are calculated analogously. Thus, the mass fluxes through the CV face are made dependent on the pressure at the grid nodes on either side of the face. This is the basic concept of the staggered variable arrangement successfully used for the calculation of incompressible flows in recent years [12].

Substituting the mass fluxes obtained from the calculated CV face velocities in the continuity equation, in general there will be a mass imbalance S_m. Flux corrections F' are needed in order to annihilate this imbalance. These corrections are based on velocity corrections, which are further related to pressure corrections p':

$$u_{1e}' = -\overline{\left(\frac{1}{a_P - \sum a_{nb}}\right)_e} A_e (p_E' - p_P'). \tag{8}$$

Thus, the continuity constraint leads to a pressure-correction equation of the final form:

$$a_P p_P' = \sum a_{nb} p_{nb}' - S_m . \tag{9}$$

The solution algorithm can now be summarized as follows:

1. Assemble the coefficients and solve the discretized momentum equations (6) using the currently available pressure and velocity fields. Normally only one iteration of the SIP

solver is carried out due to non-linearity of the equations.

2. Calculate new mass fluxes using special interpolation for the cell face velocities (7) and check the mass imbalance for each CV.

3. Assemble the coefficients and solve the pressure-correction equation (9). Apply the SIP solver until the sum of the absolute residuals is reduced by a factor of 5 or for a maximum of 10 iterations.

4. Correct the mass fluxes, the velocities and the pressure by the obtained pressure-correction.

5. Return to step 1 until the sum of the normalized absolute residuals of the momentum equations and the pressure-correction equation has fallen below a certain limit.

6. Advance the time and return to step 1 until the prescribed number of time steps has been reached.

RESULTS AND DISCUSSION

Calculation details

For the present study, all calculations were performed on a Convex C 120 computer. Approximately 40 MBytes of core were required for the 25x25x75 grid due to the general applicability of the code. The time step size varied from 0.001 for the initial phase to a maximum of 1 at the end of the simulation [5]. The convergence criterium within each time step was chosen to be 10^{-3}. Under-relaxation with $\alpha_u = 0.8$ has been used for the solution of the momentum equations, while no under-relaxation was needed for the pressure-correction equation due to employment of the SIMPLEC algorithm on an orthogonal grid [11]. The cpu-time required for simulation of the unsteady cavity flow up to t=200 was about 30 cpu-h. The degree of vectorization of the code was limited due to the recursive structure of the SIP solver [10].

Flow structure

The typical time-averaged laminar flow structure in a lid-driven cavity of Re=3200 consists of a main circulation cell and three secondary eddies (cf. Fig. 1). The primary recirculating flow is generated by the moving wall dragging the adjacent fluid. Secondary eddies are formed in the apices of the vertical and bottom boundaries as a result of frictional losses and stagnation pressure. A third secondary eddy is generated on the upstream vertical wall near the moving lid. The principal discrepancies among two- and three-dimensional simulations have been reported by Freitas and Street [5] and Iwatsu et al. [6]. In three-dimensional cavity flow the intensity and size of the primary vortex and the secondary eddies become dependent on their spanwise location and time. Fig. 3 displays the temporal variation of the vorticity contours in plane $x_3 = 0$. The circulation pattern is modified, particularly in the region of the primary vortex and near the downstream secondary eddy.

t=50

t=100

t=200

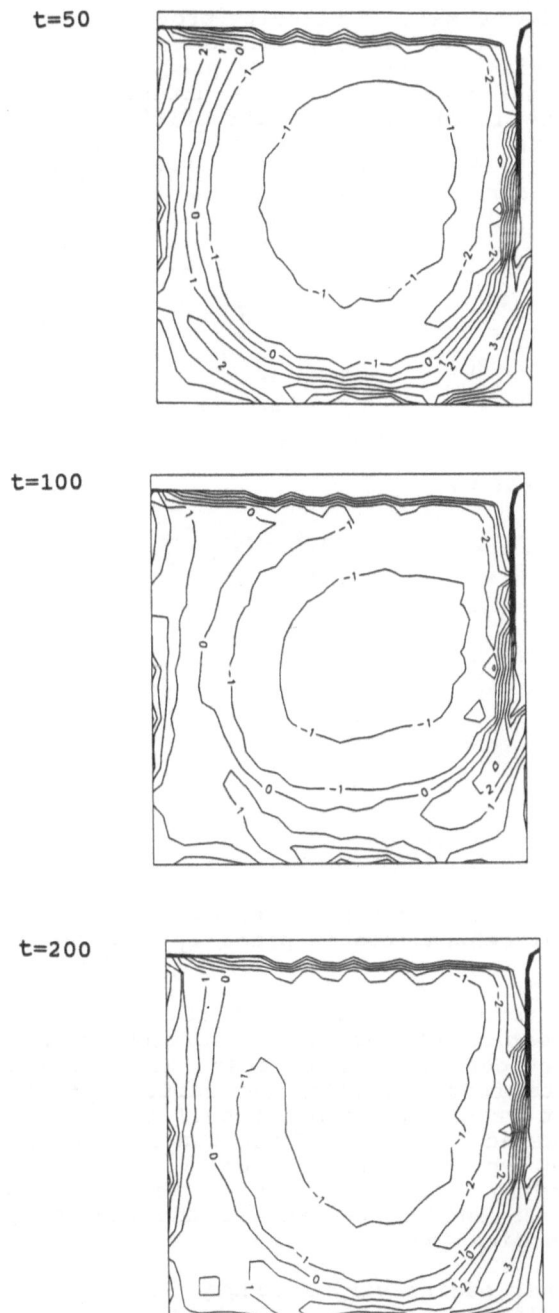

Fig. 3: Temporal variation of vorticity contours in plane $x_3=0$

Fig. 4 compares the normalized velocity profiles along the horizontal and vertical centrelines in plane $x_3=0$ at various times. The considerable time variation of the profiles near the bottom wall clearly demonstrates the unsteady nature of the flow. The development of the longitudinal Taylor-Görtler-like (TGL) vortices redistributes energy from the main flow into a direction perpendicular to it resulting in rather "weaker" flow compared to results of two-dimensional simulations [5].

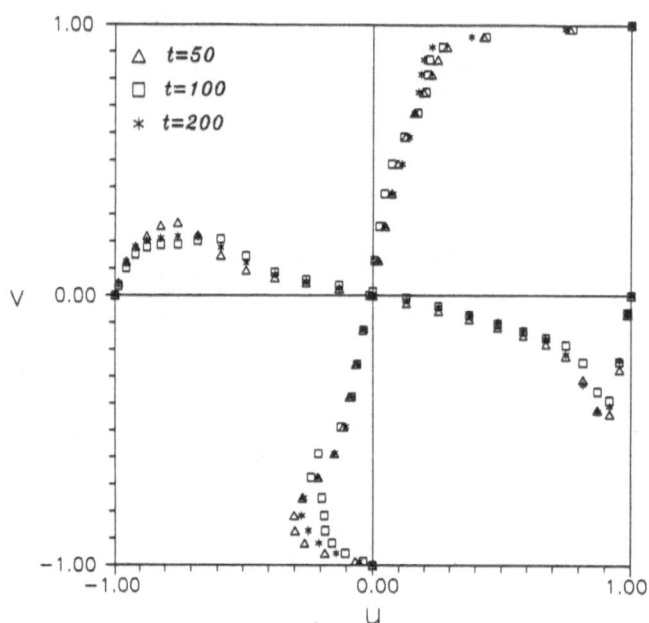

Fig. 4:   Normalized velocity profiles along the horizontal and vertical centrelines in plane $x_3=0$ at various times.

Fig. 5 shows the temporal variation of the flow field in plane $x_1=4/15$. The formation of the corner vortices and up to nine pairs of TGL vortices is clearly seen. The longitudinal TGL vortices originate from a complex interaction of the spanwise pressure gradient resulting from end-wall viscous damping and centrifugal forces acting on the fluid due to the concave shape of separation between the primary circulation cell eddy and the downstream secondary eddy. The TGL vortex formation occurs shortly before t=25 and corresponds to the development of the end-wall corner eddies [5].

The end-wall corner eddies vary strongly in size and strength. This variation relates in some degree to the size and strength of the TGL vortices. The TGL vortex pairs are asymmetric and vary significantly in pair size, in time and in space. The results obtained with the present method are in qualitatively good agreement with the experimental and numerical results reported by Street's group [1-5]. However, certain discrepancies exist in

t=5

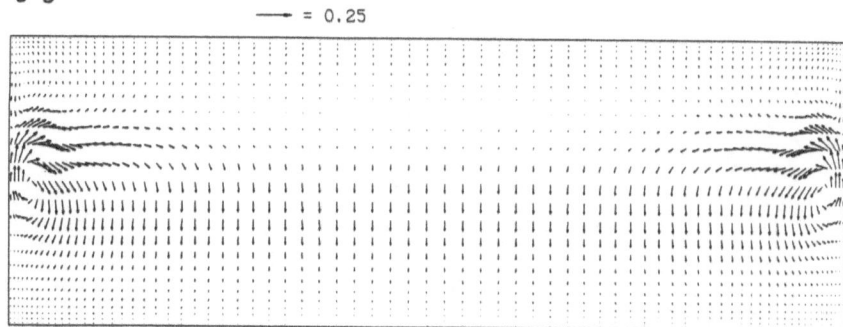

t=10

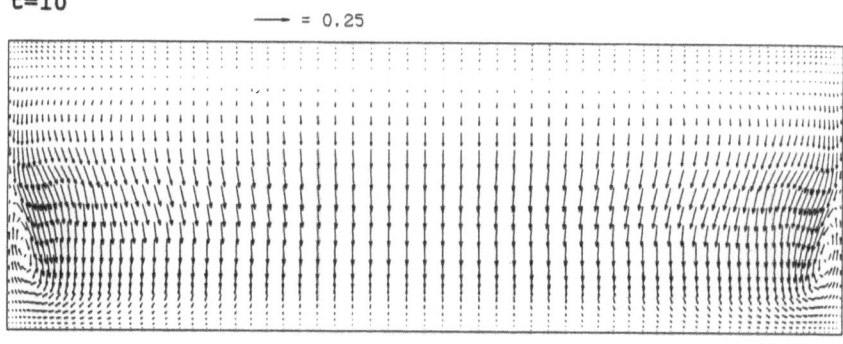

t=25

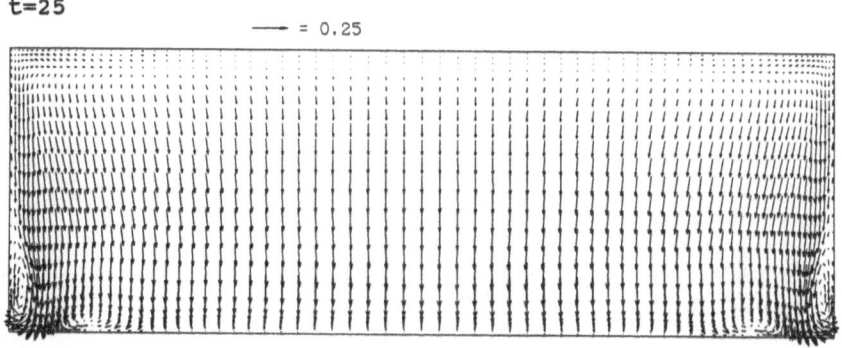

Fig. 5: Temporal variation of the flow field in plane $x_1=4/15$

t=50   → = 0.25

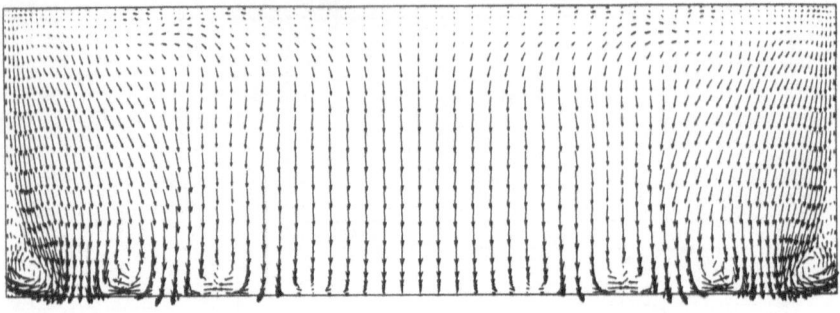

t=100   → = 0.25

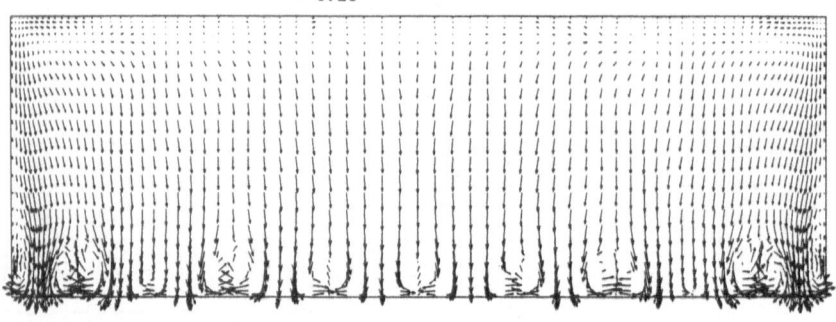

t=200   → = 0.25

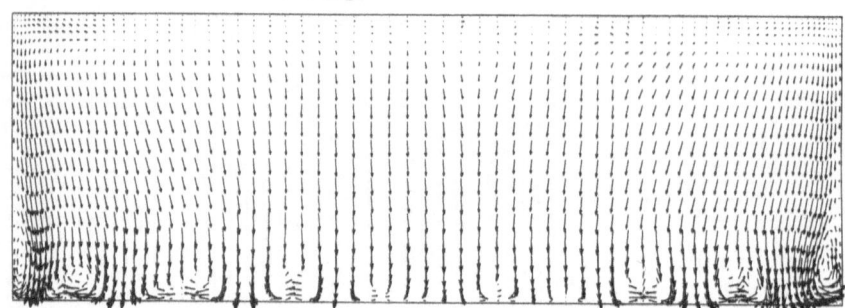

Fig. 5: (continued)

the calculated structure and location of the TGL vortices. It should be pointed out that in the present results the flow symmetry is lost between t=50 and t=100. Unfortunately, Freitas et al. [4] and Freitas and Street [5] assumed a symmetric flow and simulated only the half-cavity flow with zero flux conditions at the centre plane. However, the results reported by Iwatsu et al. [6] for the flow in a cubic cavity at a Reynolds number of Re=4000 showed a loss of symmetry at higher times. Therefore the symmetric flow assumption of Street's group requires further investigation.

CONCLUSIONS

The results presented for the three-dimensional unsteady laminar flow in a lid-driven cavity have been shown to be substantial different from two-dimensional solutions. The presence of the end-walls and the induced longitudinal Taylor-Görtler-like vortices result in a spanwise variation of the intensity and size of the primary vortex and the secondary eddies. Good agreement with previously reported experimental and numerical results has been obtained for the structure of the end-wall corner eddies and the size and number of up to nine TGL vortex pairs. In the present study the flow structure loses symmetry at higher times. Further investigation of the oscillation and meandering of the longitudinal TGL vortices is required.

ACKNOWLEDGEMENTS

This investigation was sponsored by the Deutsche Forschungsgemeinschaft through special research program SFB 278.

REFERENCES

[1]   Koseff, J.R. and R.L. Street: "Flow visualization of a recirculating flow by rheoscopic liquid and liquid crystal techniques", Experiments in Fluids, 2 (1984) pp. 57-64.

[2]   Koseff, J.R. and R.L. Street: "Visualization studies of a shear driven three-dimensional recirculating flow", J. Fluids Eng. 106 (1984) pp. 21-29.

[3]   Koseff, J.R. and R.L. Street: "The lid-driven cavity flow: a synthesis of qualitative and quantitative observations", J. Fluids Eng. 106 (1984) pp. 390-398.

[4]   Freitas, C.J., R.L. Street, A.N. Findikakis and J.R. Koseff: "Numerical simulation of three-dimensional flow in a cavity", Int. J. Num. Meth. Fluids 5 (1985) pp. 561-575.

[5]   Freitas, C.J. and R.L. Street: "Non-linear transient pheno-
      mena in a complex recirculating flow: A numerical investi-
      gation", Int. J. Num. Meth. Fluids 8 (1988) pp. 769-802.

[6]   Iwatsu, R., K. Ishii, T. Kawamura, K. Kuwahara and J.M.
      Hyun: "Numerical simulation of three-dimensional flow
      structure in a driven cavity", Fluid Dyn. Res. 5 (1989) pp.
      173-189.

[7]   Perng, C.-Y. and R.L. Street: "Three-dimensional unsteady
      flow simulations: alternative strategies for a volume-aver-
      aged calculation", Int. J. Num. Meth. Fluids 9 (1989) pp.
      341-362.

[8]   Perić, M.: "A finite volume method for the prediction of
      three-dimensional fluid flow in complex ducts", Ph.D. the-
      sis, University of London (1985).

[9]   Majumdar, S., W. Rodi and B. Schönung: "Calculation proce-
      dure for incompressible three-dimensional flows with com-
      plex boundaries", Notes on Numerical Fluid Mechanics 25:
      Finite approximations in fluid mechanics II, ed. by E.H.
      Hirschel, Vieweg, Braunschweig, 279-294 (1989).

[10] Stone, H.L.: "Iterative solution of implicit approximations
      of multidimensional partial differential equations", SIAM
      J. Num. Anal. 5 (1968) pp. 530-558.

[11] Van Doormaal, J.P. and G.D. Raithby: "Enhancement of the
      SIMPLE method for predicting incompressible fluid flows",
      Num. Heat Transfer 7 (1984) pp. 147-163.

[12] Patankar, S.V.: "Numerical heat transfer and fluid flow",
      McGraw-Hill, New York (1980).

A 3-D DRIVEN CAVITY FLOW SIMULATION
WITH PHOENICS CODE

P. MEGE

(Institut Français du Pétrole, B.P. 311
92506 Rueil-Malmaison Cédex, France)

SUMMARY

A numerical computation for 3-D unsteady incompressible flow in a driven cavity is performed using a finite volume approach via PHOENICS code on a IBM RS6000/530 work station. The Reynolds number for the simulation is 3200. The computed flow is smooth, due to the artificial viscosity introduced by the code, but significant well-formed three dimensional structures, like the Taylor-Görtler-like longitudinal vortices, appear during the time evolution.

INTRODUCTION

Physics of a three dimensionnal cavity flow at Reynolds 3200 was examined through numerical simulation essentially in Freitas *et al.* (1985) [1] . Recirculating flow, with significant secondary motion, is observed in the spanwise direction while Taylor-Görtler vortices in streamwise direction develop. Questions about the stationnarity of the solution and of the justification of a symmetry plane are posed.

CONTINUOUS PROBLEM

PHOENICS predicts the distribution of any variables, obeying Navier-Stokes conservation laws, like : the mass of the phase, the momentum of the phase via the three velocity components, the energy content via the enthalpy, and the turbulence characteristics via energy and dissipation rate. Laws of conservation of mass, momentum or energy can all be expressed in the form :

$$\frac{\partial}{\partial t}(r_i \rho_i \phi_i) + \nabla . (r_i \rho_i v_i \phi_i - r_i \Gamma_{\phi_i} \nabla \phi_i) = r_i S_{\phi_i}$$

where ϕ_i is the general conserved property for phase i, Γ_{ϕ_i} is the exchange coefficient for ϕ_i, S_{ϕ_i} is the source of ϕ_i per unit phase volume, which contains contributions such as pressure gradient in the relevant direction, gravitationnal force, interphase friction term.

BOUNDARY CONDITIONS

Phoenics expresses conditions by integration over the cells containing the points, lines, etc... ; therefore the boundary or internal conditions make contributions to source terms of the finite domain equations. Sources of mass are specified by coefficients and values associated with the pressure variable, the concept being that an inflow of mass results from an external pressure which exceeds the internal pressure. On walls, the code applies a friction (usually a log-law function) due both to the weak accuracy in space of the code and the lack of discretization points (caused by the relatively high computational cost).

DISCRETIZATION

The code uses a finite domain formulation of the differential equations for conservation. Finite volume equations are derived by integration of the differential equations over control volumes of finite size which, taken together, wholly fill the domain under consideration. The conservation laws are discretized directly in physical space. Then, mass, momentum and energy are conserved at discrete level. The internal conservation law for volume Ω can be written via the Green's formula as :

$$\frac{\partial}{\partial t} \int_{\Omega} U d\Omega + \oint_{S} \vec{F}.d\vec{S} = \int_{\Omega} Q d\Omega .$$

For each cell volume Ω_j there is an estimation of the volume and cell face areas of Ω_j, and an approximation of the fluxes at the faces :

$$\frac{\partial}{\partial t}(U_j \ \Omega_j) + \underset{\text{sides}}{\Sigma} \ (\vec{F}.\vec{S}) = Q_j \ \Omega_j .$$

There are three contrainsts on the choice of the Ω_j :

- their sum should cover the whole domain Ω,
- adjacent Ω_j may overlap if each internal surface is common to 2 volumes,
- fluxes along a cell surface have to be computed by formulas independant of the cell in which they are considered.

Staggered meshes are currently applied ; cells and nodes for velocity components are staggered relative to those of all other variables.

The arrangement is shown in the figure 1 where mass conservation is discretized on the volume (i, j), while x-momentum conservation is expressed for the volume centred on cell face, and where U velocity component are located (i+1/2, j) (Idem for y-momentum conservation on (i, j+1/2)).

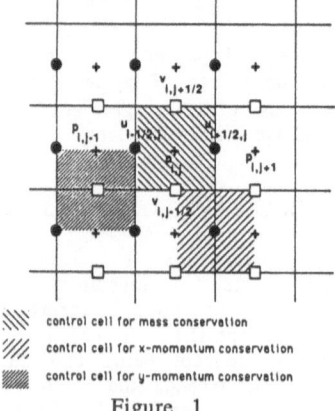

Figure 1

Integration is performed by rules concerning values of variables and values of variable gradients, prevailing at cell boundaries.

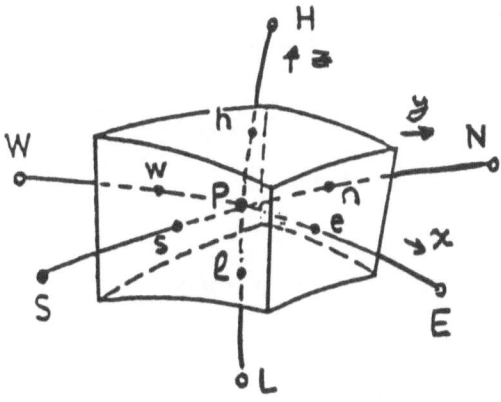

Figure 2

Integration leads to finite domain equations, having the form :

$$a_p \phi_p = a_N \phi_N + a_S \phi_S + a_E \phi_E + a_W \phi_W + a_H \phi_H + a_L \phi_L + a_T \phi_T + b .$$

The N, S, E, W, H and L are the influence of convection and diffusion terms, while T is the influence of the previous time steps The Peclet number is defined as the product between the Reynolds and the Prandtl number. If Pe < 2, a central difference scheme of second order is used, while for higher Pe, artificial viscosity is introduced by an upwind scheme of only first order.

SOURCE TERMS

In time dependent problems, a fully implicit option is used. All the source terms are taken into account on the right hand side. Source terms are used to express boundary or internal conditions, and also under-relaxation devices (false time step or linear under relaxation).
Interphase friction, interphase mass or heat transfer, chemical reactions, gravity influences or other are all represented as source terms.

THE SIMPLEST ALGORITHM

The equations for all the velocity components in a field are solved simultaneously via the momentum equations. Pressure is solved and corrected via the continuity equation. Then the convergence of the solution is based upon the respect of the mass conservation. The algorithm is [2] :

-1- initialisation with a guessed pressure p*.

-2- velocity calculation

$$a_e \, u_e^* = \Sigma \, a_{nb} \, u_{nb}^* + b + A_e \, (p_P^* - p_E^*) \, .$$

diffusion term $\quad\quad$ convection + source term $\quad\quad$ pressure gradient

-3- the pressure correction is made in order to enforce the continuity equation

$$a_p \, p_p' = \Sigma \, a_{nb} \, p_{nb}' + d.c.e.^*$$

d.c.e. = discrete continuity equation.

-4- pressure calculation

$$p = p^* + \alpha \, p'$$

α : under-relaxation factor = 0.8 .

-5- velocity calculation

$$u_e = u_e^* + d_e \, (p_P' - p_E') \, .$$

In order to facilitate the convergence, the source terms are linearized and an underrelaxation, like for the pressure, can be applied to all the components.

APPLICATION

The present application is the driven cavity at Reynolds 3200. The stretched cartesian grid is composed of 50*40*75 nodes, the size of the box is 1*1*3*L_0, the minimal size of a node is 0.005*L_0 while the dimensionless time step is 1. The challenge of our participation at this workshop is to test if it is possible to find a realistic solution. It is clear that such a discretization does not permit to capture really unsteady solutions. But we hope that the code is sufficiently consistent and stable, in order to find a solution in good qualitative agreement with others.

The computational cost on IBM RS6000/530 is 5.6 10^{-3} second per node per time step, that is 45 hours for 200 time steps.

We note the (x, y, z) components of the velocity by (U, V, W). The often used comparison is based upon the V and U velocity profile curves at the symmetry plane on the centerlines (z=0, y=0) and (z=0, x=0) respectively (figure 3). These results are relatively time-independent until t=200, and seem to be in good agreement with those of Perng & Street [3], the location and the values of the extrema are nearly identical. Figure 4 presents the variation of the pressure along the top line (z=0, y=1/2) and along the centerline (x=0, y=0). On the first curve, pressure is exactly constant except very near the walls, due to the boundary conditions imposed to the fluid on the top of the box in the streamwise direction.

A very small wave appears on the second curve, being consistent with the weak deviation of the primary vortex center around the z-axis (cf figure 5).

Figure 5 presents spatial variation of the velocity vector in a z plane. In the z=0 plane, the secondary downstream eddies are well formed, while in the left upper corner the secondary structure is not found. The comparison between the plane z=-1 (first picture) and z=1 (third picture) shows clearly the non-symmetry of the flow at this time.

Figure 6 depicts the value of the x-component of the vorticity along the z-axis (x=0, y=0) (second picture) and the z-component of the vorticity along the top of the cavity (z=0, y=1/2) (first picture). We observe that Ω_x is strongly unsteady, the values at t=200 seem to indicate the onset of some oscillation.

Time evolution of velocity vectors in the spanwise plane x=0 is shown in figure 7. First of all, at time t=50, only endwall corner eddies are generated. The flow is then really not symmetric. The TGL vortices are apparently generated near the wall and they seem to propagate towards the center of the box. In absence of all exterior excitation, the mechanism of the formation must be the endwall effect. At time t=100, 8 pairs of vortices appear, independently of the use of symmetry conditions at z=0 (the computation have been performed with or without those conditions). At t=200, we only find 6 vortex pairs with increasing height. Further investigations of the results show that the solution at t=200 enters into a regime which is difficult to be captured with our discretization. Thus, the figure 13 shows some strong oscillations of the isovalues of Ω_z in the plane y=1/2.

Figures 8 to 13 present more details for the isovalues of vorticity and pressure.

CONCLUSION

The discretization which can reasonably be used on a work station is sufficient to compute the flow and to capture the structures (like the TGL vortices) correctly during a certain time interval. Nevertheless, there must be some further instabilities (cf results at t=100 and t=200) which need a higher resolution in space and time in order to prevent the divergence of the numerical simulation. Therefore the kind of code we used is not able to compute flows with very fine structures ; the need of a high discretization would lead to astronomical computational costs. But this is not the aim of such a code : they are certainly very useful to compute steady solutions even with a certain spatial complexity.

Thanks to Université Claude Bernard (Lyon) for the loan of an IBM RS6000/530

REFERENCES

[1] Freitas C.J., Street R.L., Findikakis A.N., Koseff J.R. : "Numerical simulation of 3-D flow in a cavity". Int.J.Numer.Methods.Fluids., Vol 5, 561-575, 1985.

[2] Patankar S.V. : "Numerical heat transfer and fluid flow".Series in Computational Methods in Mechanics and Thermal Sciences, Hemisphere Publishing Corporation, 1980.

[3] Perng C.Y., Street R.L. : "3-D unsteady flow simulations : alternative strategies for a volume-averaged calculation". Int.J.Numer.Methods.Fluids., Vol.9, 341-362, 1989.

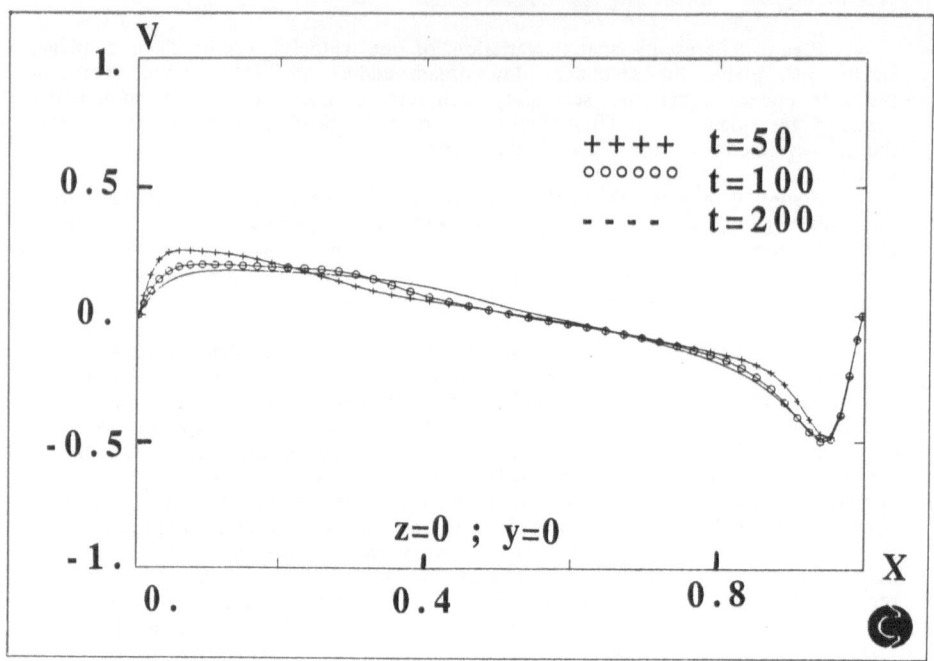

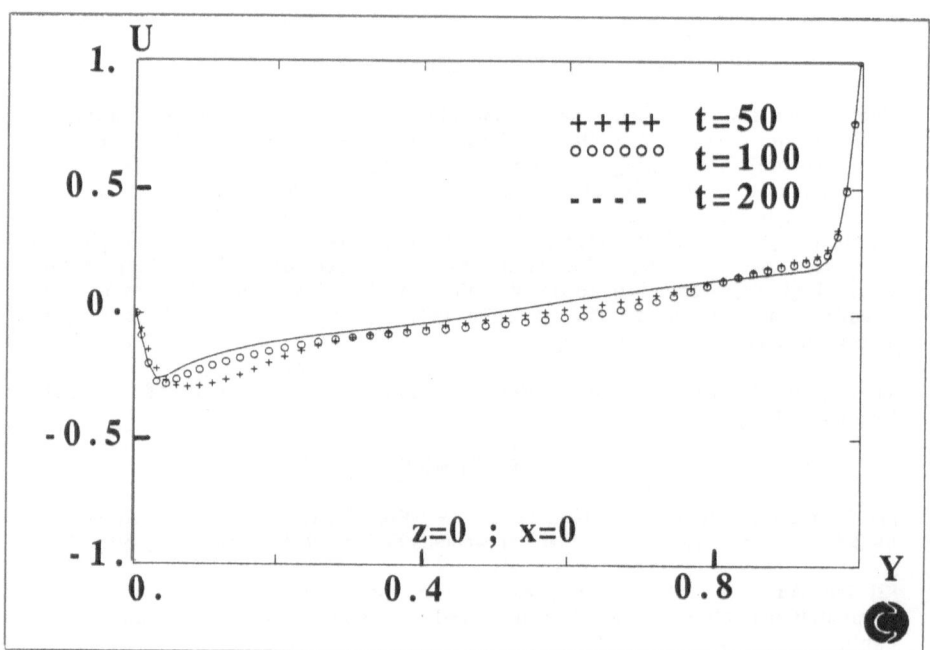

Figure 3

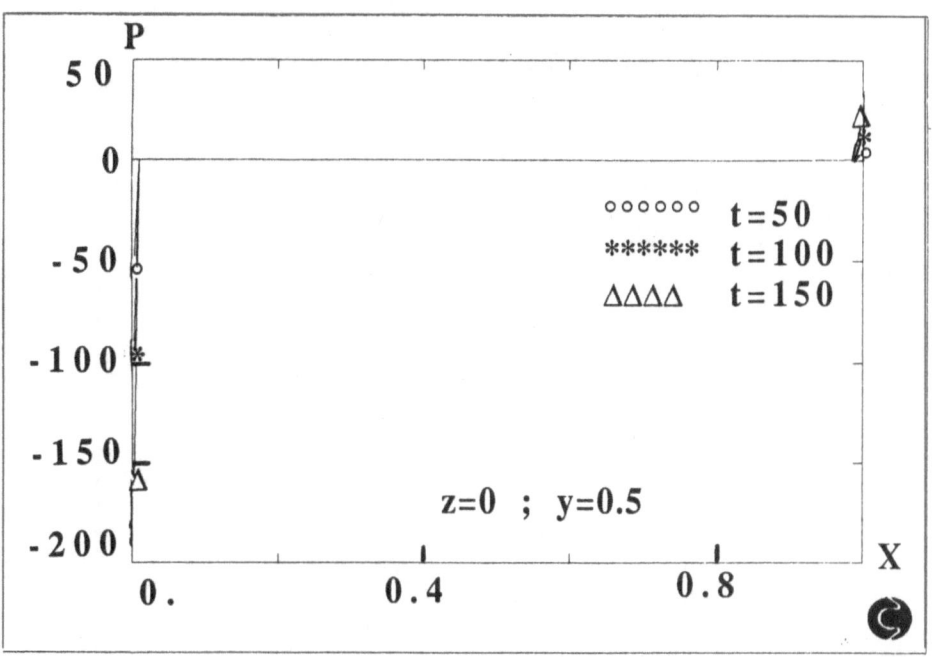

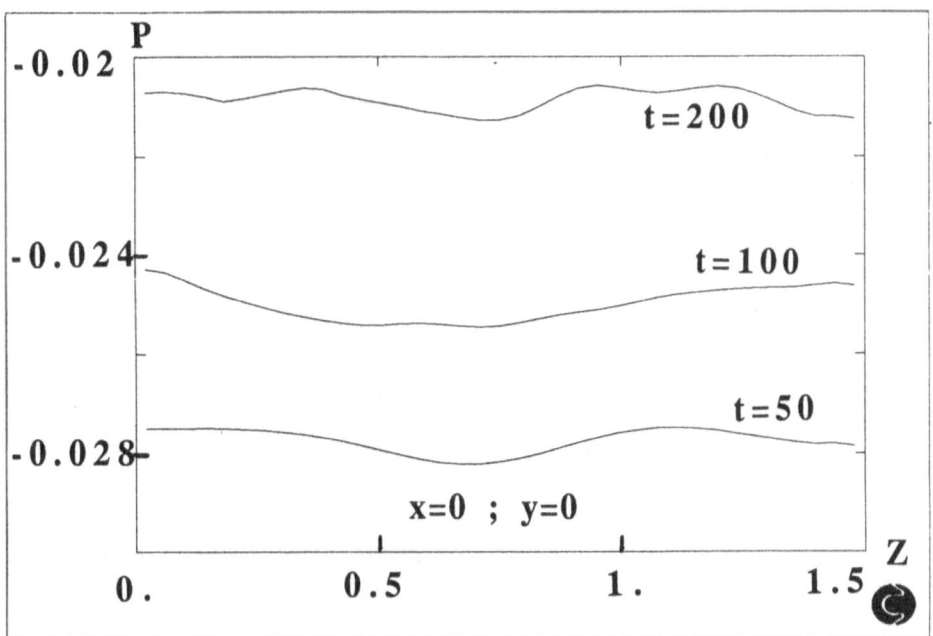

Figure 4

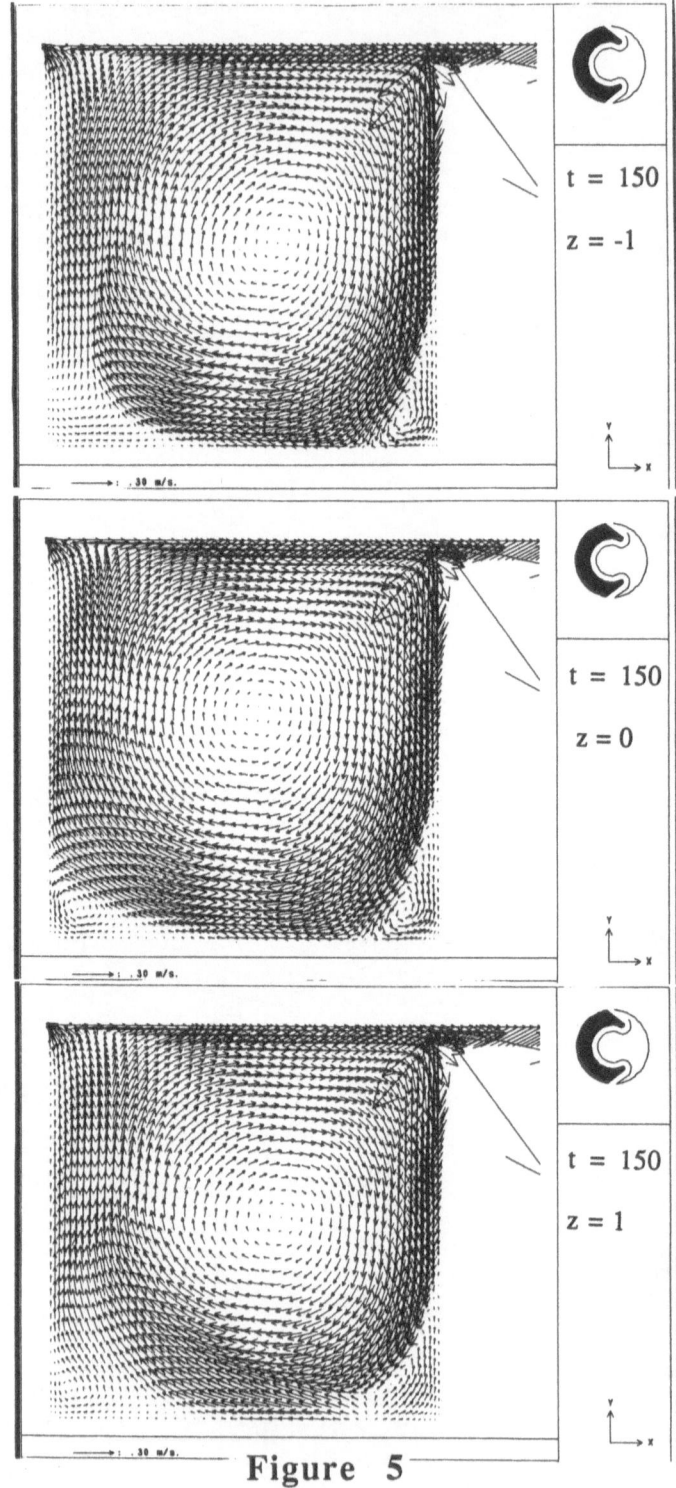

t = 150
z = -1

t = 150
z = 0

t = 150
z = 1

Figure 5

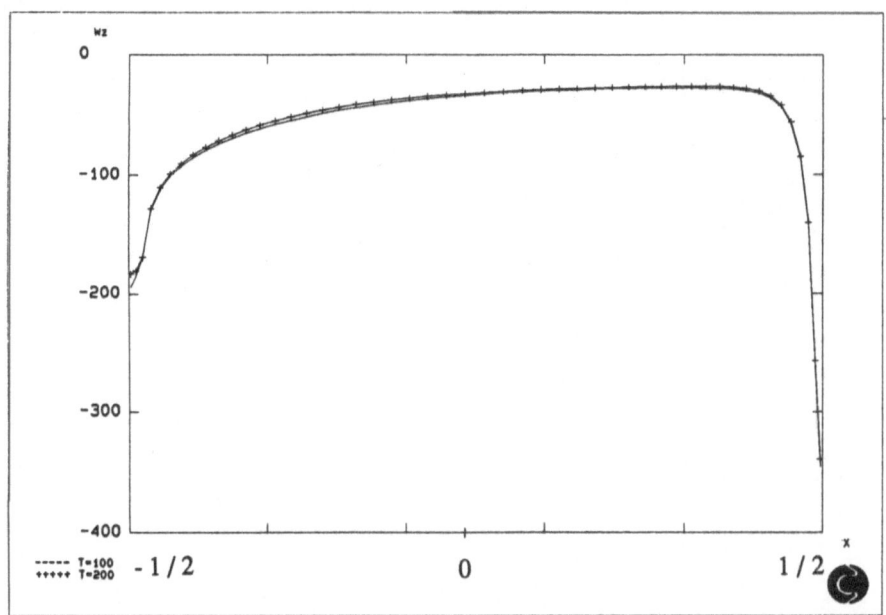

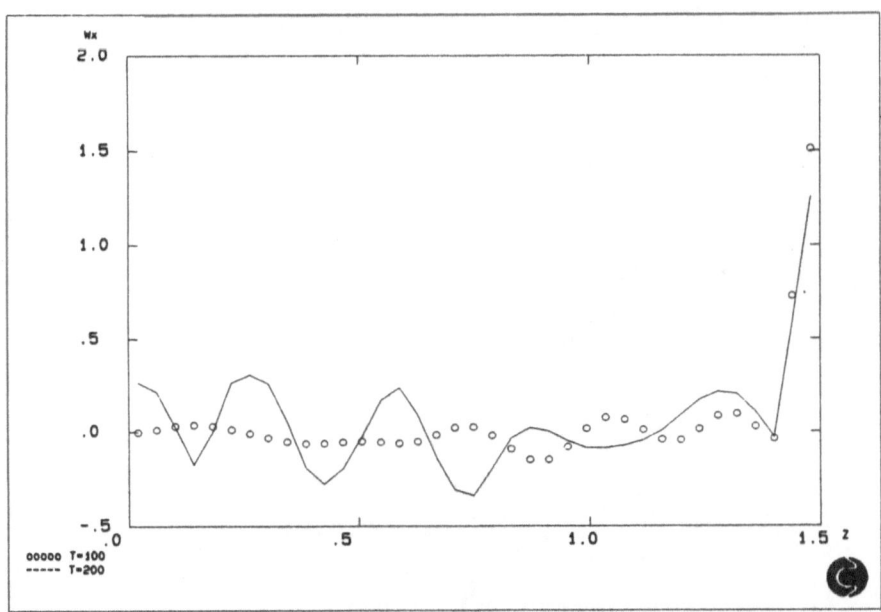

Figure 6

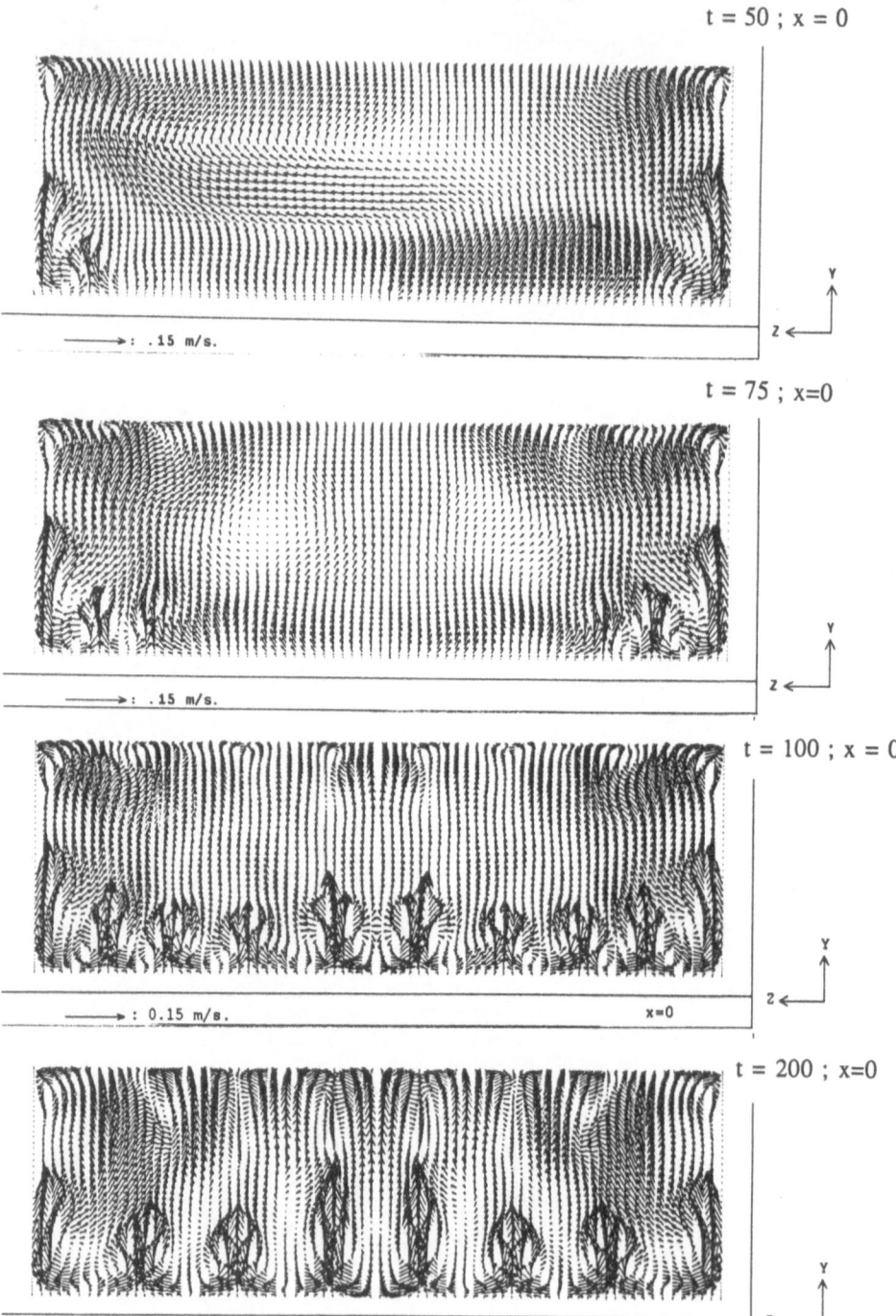

t = 50 ; x = 0

→ : .15 m/s.

t = 75 ; x=0

→ : .15 m/s.

t = 100 ; x = 0

→ : 0.15 m/s.   x=0

t = 200 ; x=0

→ : 0.15 m/s.

Figure 7

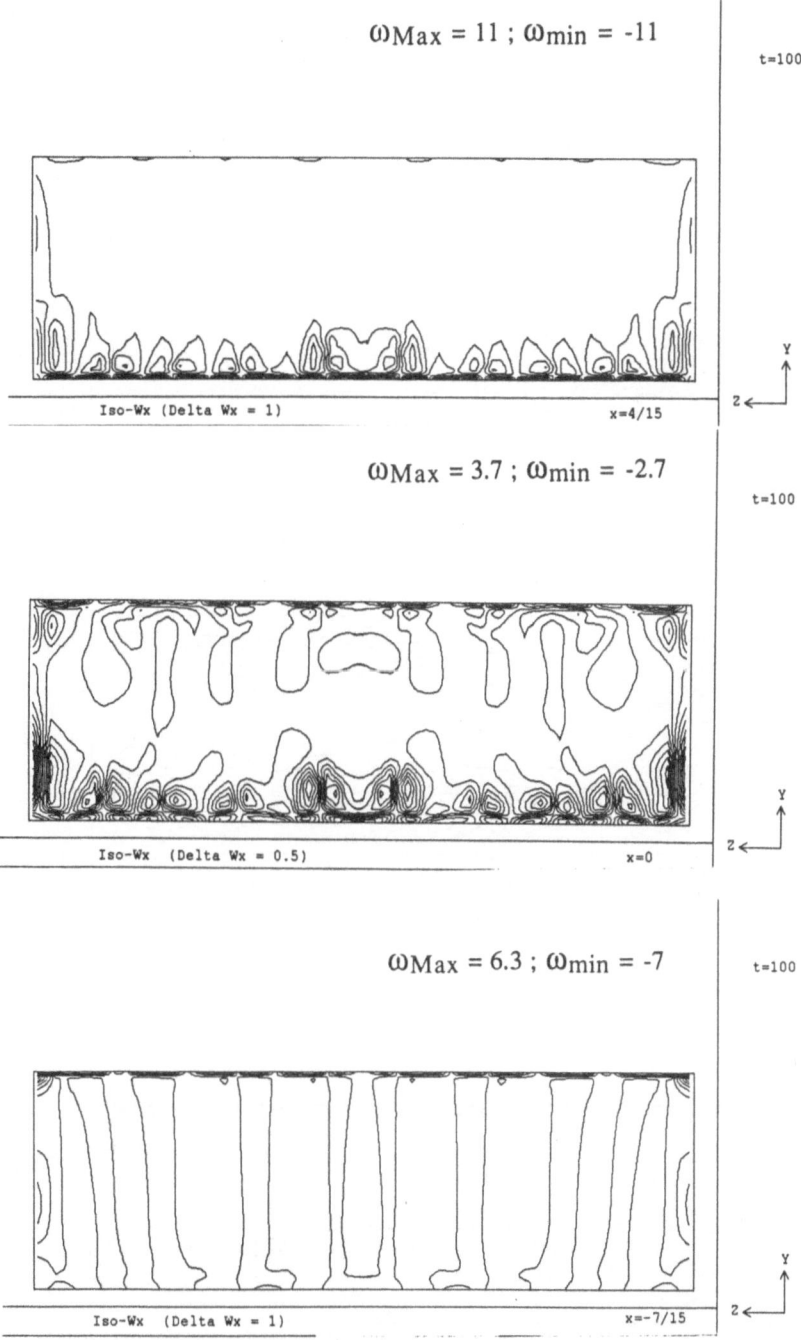

$\omega\text{Max} = 11$; $\omega\text{min} = -11$

t=100

Y

Z

Iso-Wx (Delta Wx = 1)   x=4/15

$\omega\text{Max} = 3.7$; $\omega\text{min} = -2.7$

t=100

Y

Z

Iso-Wx (Delta Wx = 0.5)   x=0

$\omega\text{Max} = 6.3$; $\omega\text{min} = -7$

t=100

Y

Z

Iso-Wx (Delta Wx = 1)   x=-7/15

Figure 8

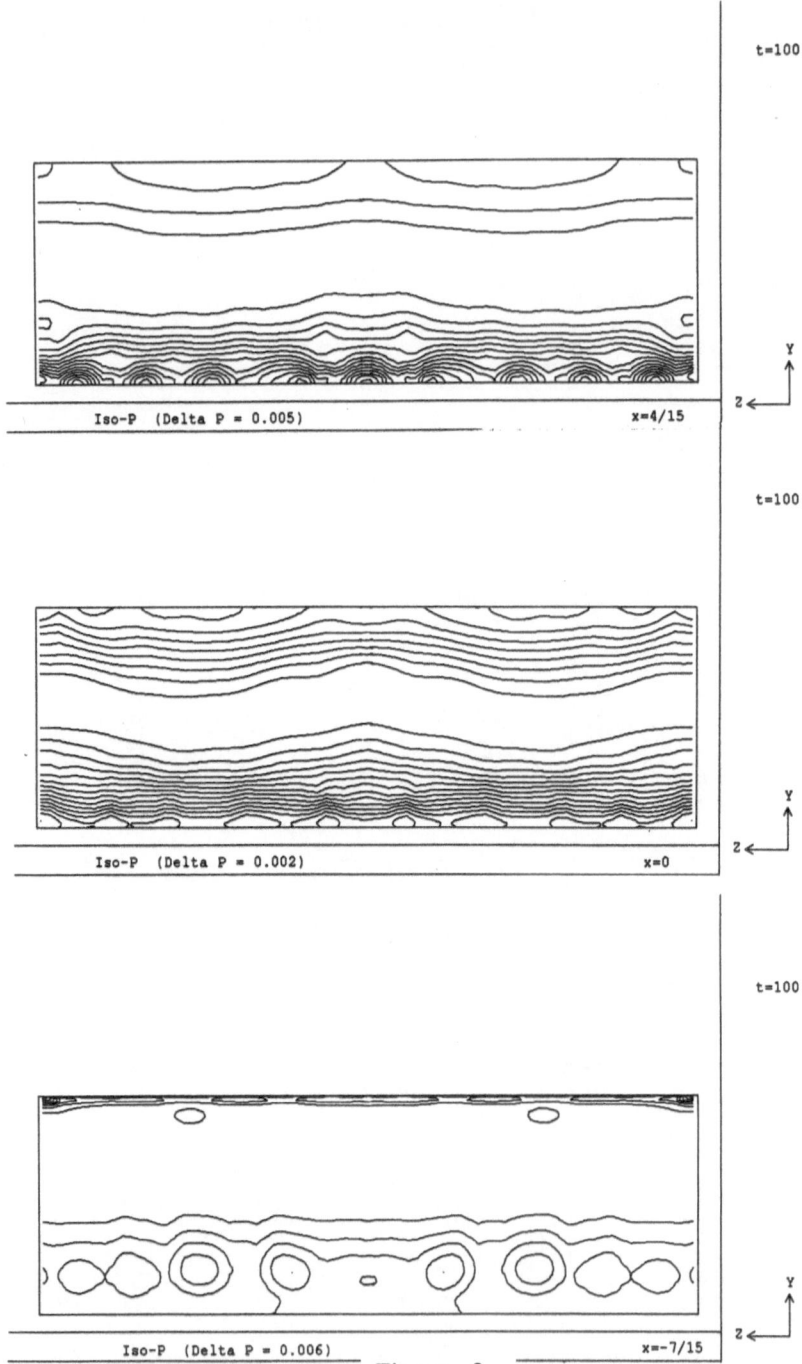

Figure 9

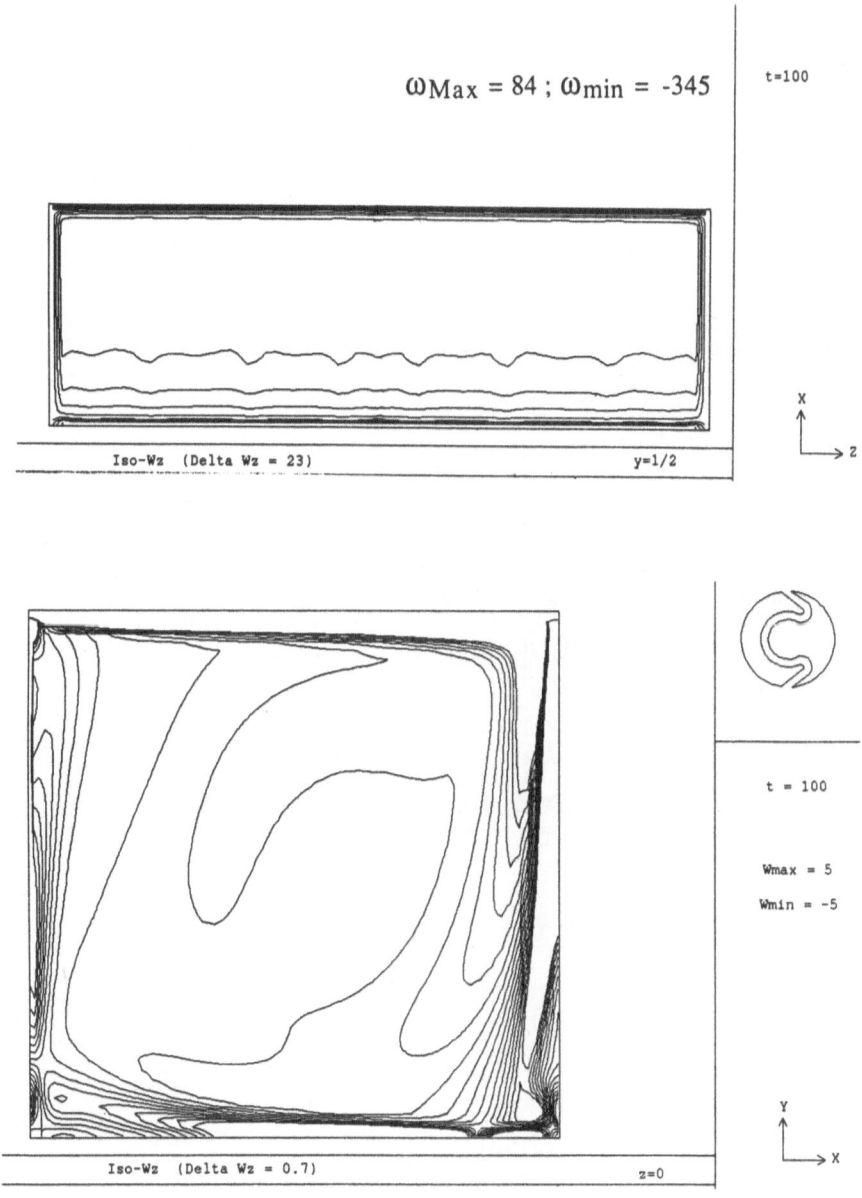

$\omega \text{Max} = 84 \ ; \ \omega \text{min} = -345$

t=100

Iso-Wz   (Delta Wz = 23)   y=1/2

X
↑
→ Z

t = 100

Wmax = 5

Wmin = -5

Iso-Wz   (Delta Wz = 0.7)   z=0

Y
↑
→ X

Figure   10

103

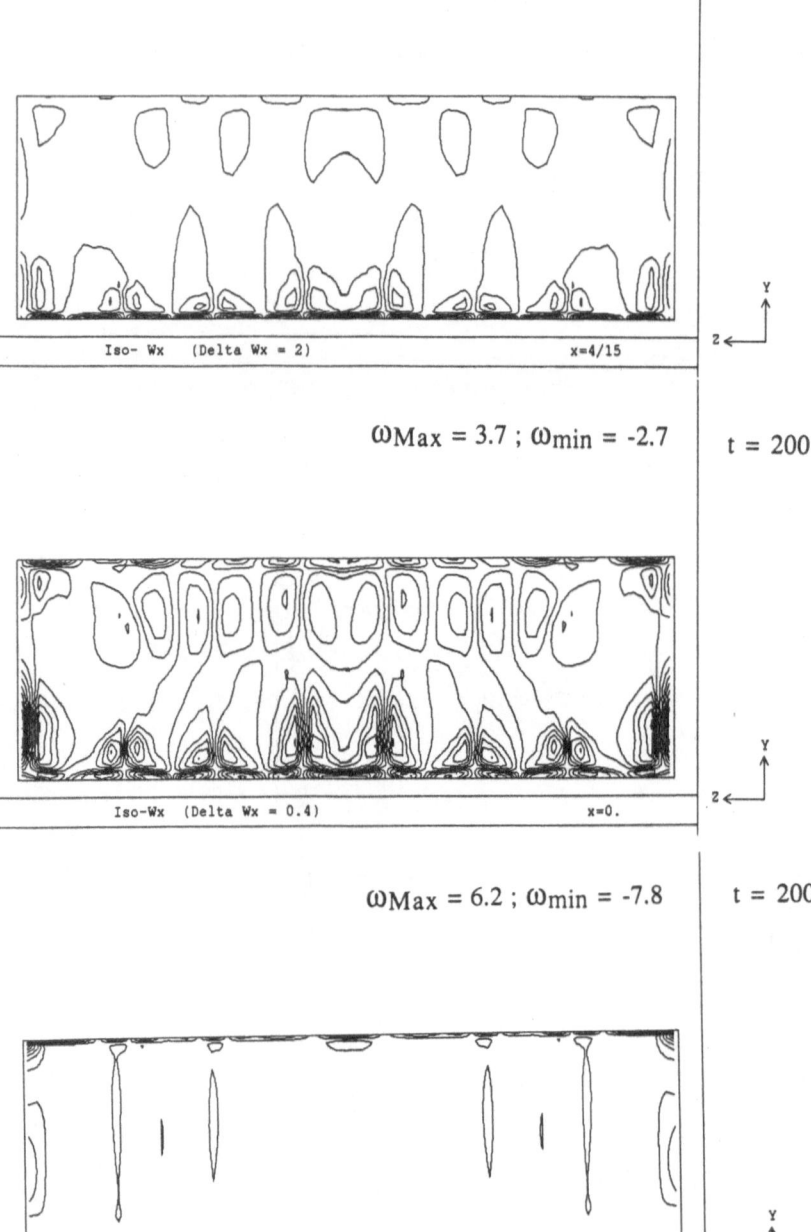

ωMax = 15 ; ωmin = -12   t = 200

Iso- Wx   (Delta Wx = 2)   x=4/15

ωMax = 3.7 ; ωmin = -2.7   t = 200

Iso-Wx (Delta Wx = 0.4)   x=0.

ωMax = 6.2 ; ωmin = -7.8   t = 200

Iso-Wx (Delta Wx = 1)   x=-7/15

Figure 11

104

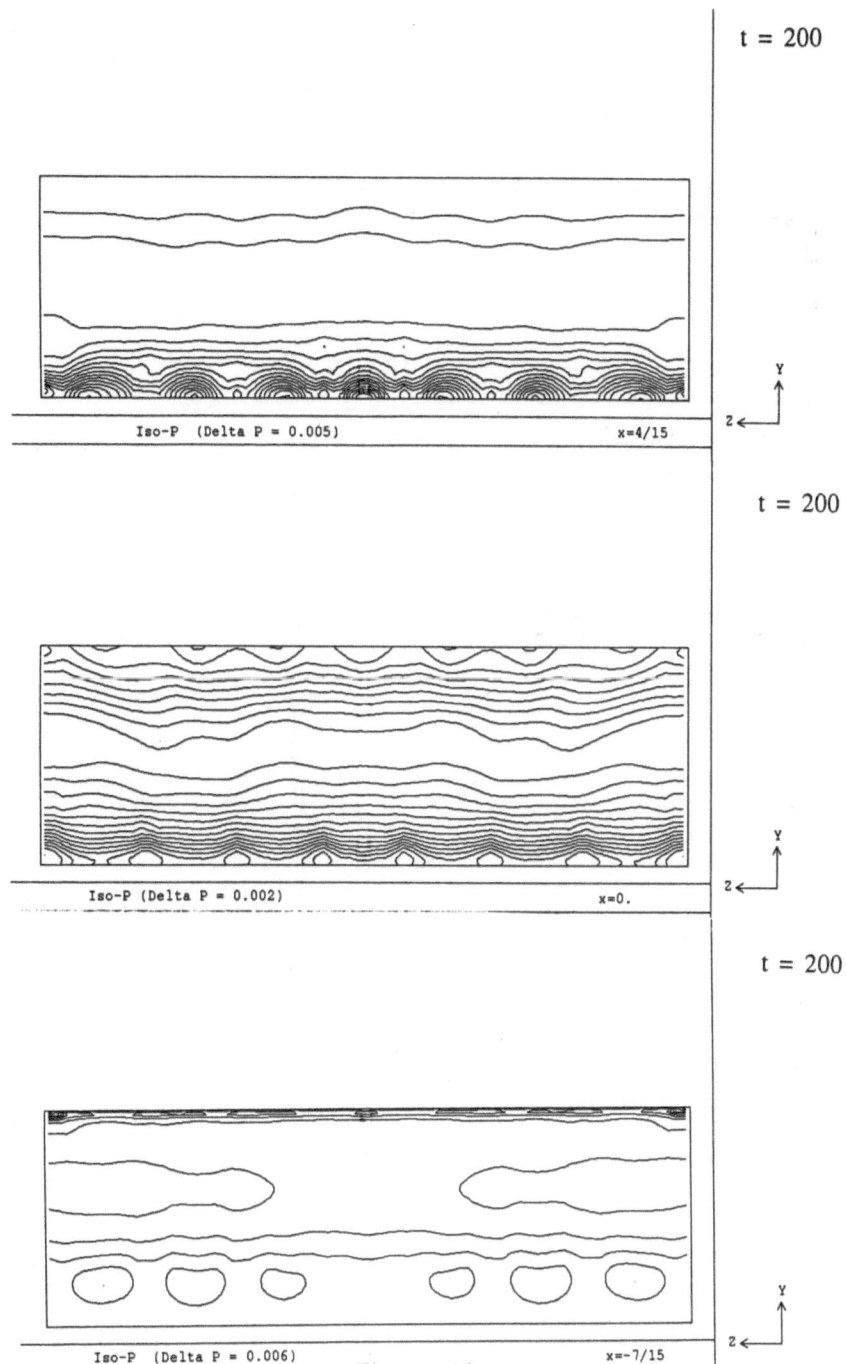

t = 200

Iso-P (Delta P = 0.005)   x=4/15

t = 200

Iso-P (Delta P = 0.002)   x=0.

t = 200

Iso-P (Delta P = 0.006)   x=-7/15

Figure 12

105

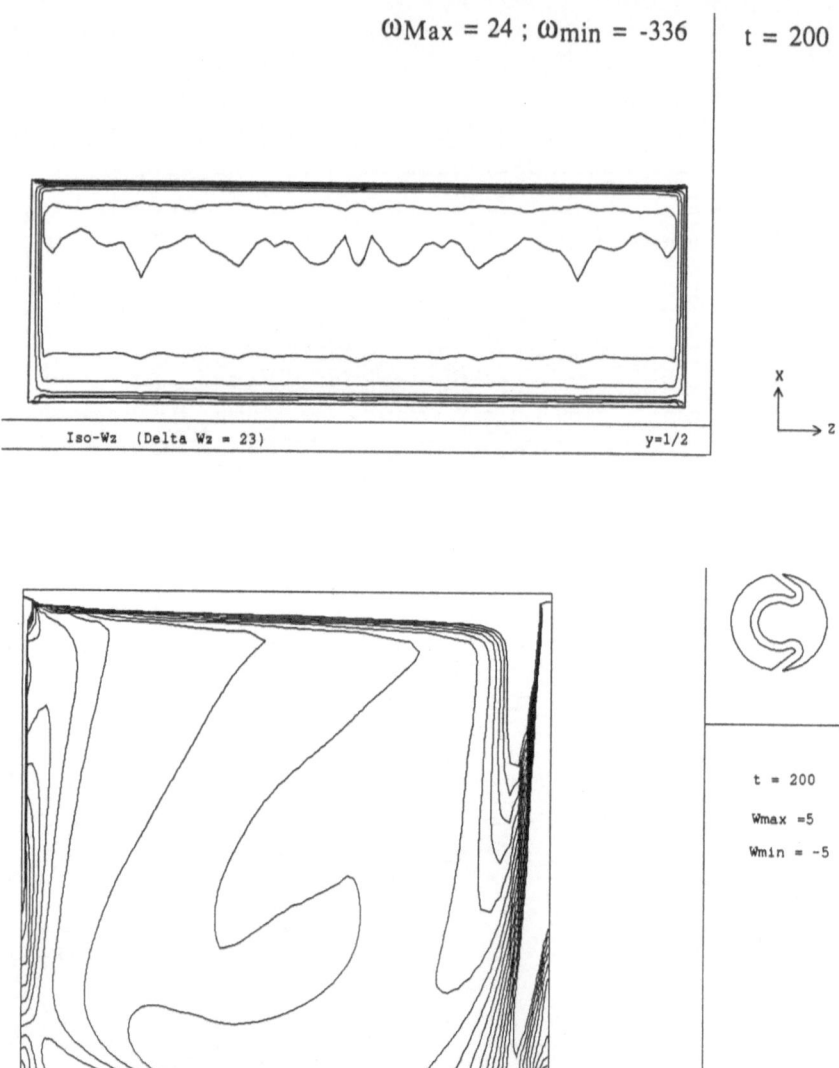

ωMax = 24 ; ωmin = -336   t = 200

Iso-Wz (Delta Wz = 23)   y=1/2

X
↑
└─→ Z

t = 200

Wmax =5

Wmin = -5

Iso-Wz (Delta Wz = 0.7)   z=0

Y
↑
└─→ X

Figure 13

Multigrid and ADI techniques to solve unsteady 3D viscous flow in velocity-vorticity formulation

D. TROMEUR-DERVOUT (*), L. TA PHUOC (**)

(*) ONERA, Parallel Computing Division , BP 72, 92322 Chatillon Cedex, FRANCE
(**) LIMSI/CNRS and ONERA, BP 133, 91403 Orsay Cedex, FRANCE

Summary

Multigrid and ADI methods were used to solve unsteady 3D Navier-Stokes equations in Velocity-Vorticity formulation, for the lid-driven cavity test case of spanwise aspect ratio equal to 3:1, at a Reynolds number Re equal to 3200. Results are given for the characteristic times T=50,100,200.

I. Introduction

Navier-Stokes equations by finite differences methods in formulation $(\vec{V}, \vec{\omega})$ are solved. An Alternating Direction Implicit Method was used to solve the transport equation associated to the vorticity vector and a Multigrid method was used to solve the elliptic problem associated to the velocity vector. In part 2, one describes the velocity-vorticity formulation while part 3 deals with the algorithm description. The lid driven cavity test case results are discussed in part 4. Conclusion and future works are presented in part 5.

II. Formulation

The Velocity-Vorticity formulation, mathematically equivalent to the primitive variables Velocity-Pressure formulation as demonstrated in [1] is chosen for this fluid flow simulation. This formulation leads to a more natural decoupling of the governing equation than in the V-P formulation, by separating the spin dynamic of a fluid particle from its translation kinematics [2][3]. It avoids the difficult problem of pressure-velocity coupling and gives vorticity directly. Its main advantage is to be sensitive to non inertial effects only by the mean of initial and boundary conditions [4].So it would be well adapted to the external motion. In return pressure must be computed by a post-treatment, and six equations must be solved.

The 3D Navier-Stokes equations are written in conservative form in a domain Ω where $\vec{V}$ and $\vec{\omega}$ denote the velocity and the vorticity respectively:

107

$$\frac{\partial \vec{\omega}}{\partial t} - \vec{\nabla} \times (\vec{V} \times \vec{\omega}) = \frac{1}{Re}\nabla^2 \vec{\omega} \quad in \ \Omega \tag{1}$$

$$\vec{\nabla}(\vec{\nabla}.\vec{V}) = \vec{\nabla} \times \vec{\omega} + \nabla^2 \vec{V} = 0 \quad in \ \Omega \tag{2}$$

$$\vec{\nabla}.\vec{\omega} = 0 \quad in \ \Omega \tag{3}$$

$$\vec{\nabla}.\vec{V} = 0 \quad in \ \Omega \tag{4}$$

$$\vec{\omega}.\vec{n} = (\vec{\nabla} \times \vec{V}).\vec{n} \quad on \ \partial\Omega \tag{5}$$

$$\vec{V} = \vec{V}_{|\partial\Omega} \quad on \ \partial\Omega . \tag{6}$$

All these equations overdetermine the problem. Only equations (1),(2), and (6) are solved , the other ones are assumed to be implicitly verified and are checked numerically in the calculation.

Numerical approximations

The six scalar equations are approximated in space by second order centered finite difference schemes using a regular non staggered mesh . The temporal discretization is obtained by an ADI scheme providing a second order time accurate scheme for the vorticity transport equation.

Boundary conditions

No slip boundary conditions are imposed for velocity while vorticity definition, $\vec{curl V}$, is used to define its boundary conditions.
In the cavity $[-\frac{1}{2}, \frac{1}{2}]^2 \times [-\frac{3}{2}, \frac{3}{2}]$ driven problem , we have:

for $x_0 = \frac{\pm 1}{2}$: $\omega_1(x_0, y, z) = 0$; $\omega_2(x_0, y, z) = -\frac{\partial V_3}{\partial x}$; $\omega_3(x_0, y, z) = \frac{\partial V_2}{\partial x}$

for $y_0 = \frac{\pm 1}{2}$: $\omega_1(x, y_0, z) = \frac{\partial V_3}{\partial y}$; $\omega_2(x, y_0, z) = 0$; $\omega_3(x, y_0, z) = -\frac{\partial V_1}{\partial y}$

for $z_0 = \frac{\pm 3}{2}$: $\omega_1(x, y, z_0) = -\frac{\partial V_2}{\partial z}$; $\omega_2(x, y, z_0) = \frac{\partial V_1}{\partial z}$; $\omega_3(x, y, z_0) = 0$.

Due to the fact that the finite difference discretization is chosen, the difficult problem of vorticity definition at the cavity corners is avoided.

III. Algorithm description

One time step of the algorithm consists in two stages: first solve the solution of the three vorticity equations, then the solution of the three velocity equations. We used at each stage the necessary updated variables.

The solution of vorticity equation

A fractional step ADI method with a stabilization correction [5] is adopted to solve the vorticity equations. The ADI method chosen for the resolution of the equation: $\frac{\partial \omega}{\partial t} = \mathbf{A}\omega - \Phi$ is based on the splitting of the operator $\mathbf{A} = (L_x + L_y + L_z)$.
The fractional step algorithm for computing ω at time $k+1$, is defined as follows:

$$(L_x - \frac{2}{\Delta t})\omega^{k+\frac{1}{3}} = -(L_x + 2L_y + 2L_z + \frac{2}{\Delta t})\omega^k + 2\Phi \tag{7}$$

$$(L_y - \frac{2}{\Delta t})\omega^{k+\frac{2}{3}} = L_y\omega^k - \frac{2}{\Delta t}\omega^{k+\frac{1}{3}} \tag{8}$$

$$(L_z - \frac{2}{\Delta t})\omega^{k+1} = L_z\omega^k - \frac{2}{\Delta t}\omega^{k+\frac{2}{3}}. \tag{9}$$

For the first vorticity component L_x, L_y, L_z take the following form:

$$L_x\omega_1 = \frac{1}{Re}\frac{\partial^2\omega_1}{\partial x^2}$$

$$L_y\omega_1 = \frac{1}{Re}\frac{\partial^2\omega_1}{\partial y^2} - \frac{\partial(V_2\omega_1)}{\partial y}$$

$$L_z\omega_1 = \frac{1}{Re}\frac{\partial^2\omega_1}{\partial z^2} - \frac{\partial(V_3\omega_1)}{\partial z}.$$

For the second vorticity component they are written as :

$$L_x\omega_2 = \frac{1}{Re}\frac{\partial^2\omega_2}{\partial x^2} - \frac{\partial(V_1\omega_2)}{\partial x}$$

$$L_y\omega_2 = \frac{1}{Re}\frac{\partial^2\omega_2}{\partial y^2}$$

$$L_z\omega_2 = \frac{1}{Re}\frac{\partial^2\omega_2}{\partial z^2} - \frac{\partial(V_3\omega_2)}{\partial z}.$$

Similar equations hold for the third component.
With the discretizations adopted ,the operators L_x, L_y, L_z, are tridiagonal, and the solution of each fractional step is simply done by (LU) factorization.

The solution of velocity equation

A linear Multigrid method [6] is chosen and has been implemented to solve the elliptic velocity equations $\nabla^2\vec{V}^{k+1} = -\vec{\nabla} \times \vec{\omega}^k$ at time $k+1$.
Standard multigrid methods work with a series of coaser grid (H), each obtained by eliminating every other node of the previous grid (h). The error equation for the fine grid is then projected to the coarse grid at every second node. The coarse grid equation is solved approximately, and the error is interpolated back to the fine grid and added to the solution there. Recursive application of this procedure

defines complete multigrid procedure.

A linear V-cycle is chosen using the fact that updated variable provides a good first approximation of the next time step solution, so F-cycle procedure is not necessary. The smoother used , is an implicit by plane Gauss-Seidel sweep alternated in the three spatial directions, in order to vectorize internal loops. Three pre-smoothing , three post-smoothing and sixty basis method iterations on coarse grid were operated. We used three grid-level and seven V-cycles. The fine-to-coarse grid transfer consists in a full coarsening procedure, keeping one node over two in each direction. The coarse-to-fine grid transfer is a trilinear interpolation.

Pressure calculation

Pressure fields are calculated when it's desired, by the method developed in [7].

IV. The lid driven cavity simulation

This method is applied to the lid-driven cavity test case with Reynolds number equal to 3200.This problem has been studied experimentally by Street & al [8][9], and some experimental data and visualization have been given.

The computational domain is $\{(x, y, z) \in [-\frac{1}{2}, \frac{1}{2}]^2 \times [-\frac{3}{2}, \frac{3}{2}]\}$ with :

- Initial conditions $V = \omega = 0$ for $t \leq 0$

- Boundary conditions for $t \geq 0$:
  $V = (0, 0, 0)$ for $(x = \pm\frac{1}{2}, y = -\frac{1}{2}, z = \pm\frac{1}{2})$
  $V = (1, 0, 0)$ for $(y = +\frac{1}{2})$.

Time step and space grid

The full cavity is represented by a grid of $(41 \times 41 \times 81)$ non staggered regular nodes, with $\Delta x = \Delta y = 0.025$, $\Delta z = .0375$. The time step is chosen in relation with the space step in order to have a CFL number of 2, $\Delta t = 0.05$.

Computation cost

This fluid motion simulation has been done on the Cray-YMP computer. The code implementation could support a stretched mesh, so it requires more computations than if it uses only a regular mesh. Memory requirement is 2.5 Mega-words (64-bits) for a grid of $41 \times 41 \times 81$ nodes. This code is running at 140 Mflops on one processor of Cray-YMP in multi-users class. Average computing cost is $3.34s$ per time step, so nearly $2.5 \times 10^{-5}s$ per node and per time step. Time T=200 has been reached in about 3 hours 40 minutes . The resolution of the velocity equations took about 90 percent of time execution with the chosen multigrid parameters. The pressure calculation took about 2s per time step but this time could be reduce by using enhance solver as multigrid. At this point no special effort was made on vectorization (no loop unrolling , no loop compress).

Numerical results

Velocity equations are solved with a residual of 10^{-11}. Velocity divergence is equal about 10^{-4} during total execution while Vorticity divergence is equal to 10^{-2} . As both divergences remain constant with time, the continuity equation is satisfied. The residual of the elliptic velocity equation is the gradient of the velocity divergence. The initial value of the divergence is due to the truncation error and to the non staggered grid used. The relation $\vec{curl}\vec{V} = \vec{\omega}$ is also checked numerically in the field and is satisfied with a residual less than 10^{-3} .

Figures 1.a and 1.b show the $U(y)$ and $V(x)$ curves. The solution obtained is in the area of the experiment of Freitas eventhough these results have a less amplitude. Two extrema can be seen in the downstream area. The distance between these two extrema varies with the time.

Figure 2 gives the fields in the symmetry plane z=0 at T=200. For the velocity field we do not retrieve the well known 2D structure.

Figure 3 gives the fields in the planes z=0.0375,z=$\frac{3}{4}$,z=1.4625 at T=200. The 2D structure is retrievied at plane z=0.0375 near the symmetry plane while it is not present at plane z=$\frac{3}{4}$. the center of the main eddy move with the location. Indeed , the tridimensional effects are present.

Figure 4 represents ω_x iso-values at plane $x = \frac{4}{15}$. The Taylor Gortler like vortices formation can be seen. At T=50 there are nine pairs of vortices. At T=100, some vortice structures seem to be of less amplitude and their number seems to be ten pairs. Two of these pairs seem to be little structures of the flow. At T=200 we have nine pairs of vortices fully formed and structured . This vortices location moves with the time. So the flow must be considered as unsteady .

Figure 5 shows the flow structure (velocity fields) corresponding to the plane $x = \frac{4}{15}$ at time equal to 50,100 and 200. The TGL vortices are present again. So the formulation has a good capability to represent vortices.

Figure 6 shows the pressure field for the same plane. The pressure structure seems to correspond with the vorticity structure.

The structure of the flow remains symmetric in z direction through the time.

V. Conclusion and future work

ADI and multigrid methods were used to solve unsteady Navier-Stokes equations in Velocity-Vorticity formulation. In spite of the low number of discretization nodes , our method has provided the main structures of the flow, (the main and secondary eddy , Taylor Gortler like vortices). Some structures of the flow which differ from the 2D results, prove the presence of the tridimensional effects. As the TGL vortices locations move with the time the flow is unsteady. The use of CFL number greater than one is permited. Hence the Velocity-Vorticity formulation seems to be well adapted for this kind of flow.

In future works, we shall focus in two directions. On one hand, we plan to solve both velocity and vorticity equations by multigrid techniques. This allows for the use of non uniform mesh, and avoid the decoupling in resolution step. On the second hand, a parallel version of code on a 128-node intel-iPSCi860 is under development.

Acknowledgement- This work was supported by the DRET of the French Ministry of Defense, under grant DRET6 $n°$ 8934001

References

[1] O. DAUBE , J.L. GUERMOND, A. SELLIER , " *On the velocity-vorticity formulation of Navier-Stokes equations in incompressible flow". C.R. Acad. Sci., t.313 serie 2 n°4 ,p. 377-382 ,1991.*

[2] GATSKI , *"Incompressible fluid flow computations using the vorticity-velocity formulation". Appl. Num. Math. vol 7 n°3 ,p. 227-240, 1991.*

[3] G.A. OSSWALD, K.N. GHIA, U. GHIA, ' *A direct algorithm for the solution of incompressible three-dimensional unsteady Navier-Stokes equations". AIAA 8th Computational Fluid Dynamic Conference,n° 87-1139, 1987.*

[4] C. SPEZIALE , *"On the advantages of the vorticity-velocity formulation of the equations of fluid dynamics". J. Comp. Phys. 73, p. 476-480 ,1987.*

[5] J. DOUGLAS Jr, " *Alternating direction methods for three space variables". Num. Math. 4 , p. 41-63, 1962.*

[6] A. BRANDT *"A guide to multigrid development". Lecture Notes in Mathematic', n° 560,p. 220-312 , 1981.*

[7] T-H LE, Y. MORCHOISNE, *"Treatment of pressure in incompressible viscous flow calculation". C.R. Acad. Sci. Paris, t 312, Serie II, p. 1071-1076, 1991.*

[8] J.R. KOSEFF, R.L. STREET, *"Visualization studies of a shear driven three-dimensional recirculating flow". J. Fluid Eng., Vol. 106, p. 21-29, 1984.*

[9] H.S RHEE, J.R. KOSEFF and R.L. STREET, *"Flow visualization of a recirculating flow by rheoscopic liquid and liquid crystal techniques". Exp. Fluids 2, p. 57-64, 1984.*

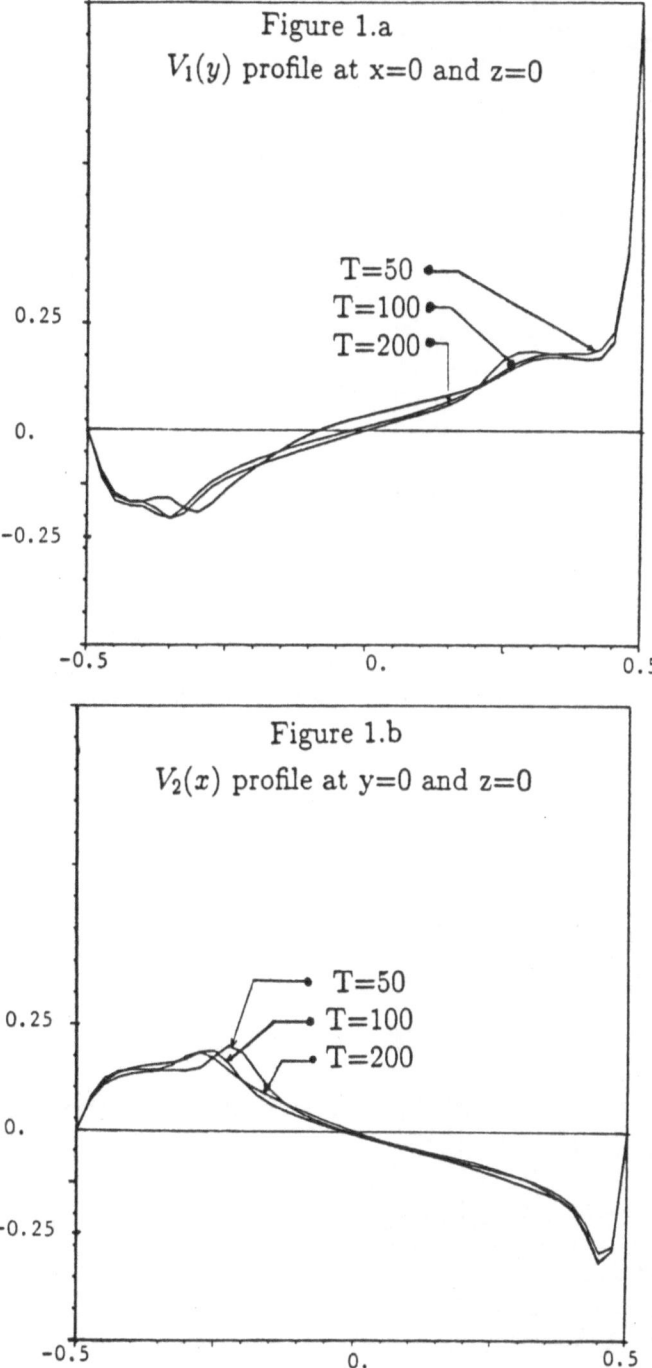

Figure 1.a
$V_1(y)$ profile at x=0 and z=0

T=50
T=100
T=200

Figure 1.b
$V_2(x)$ profile at y=0 and z=0

T=50
T=100
T=200

113

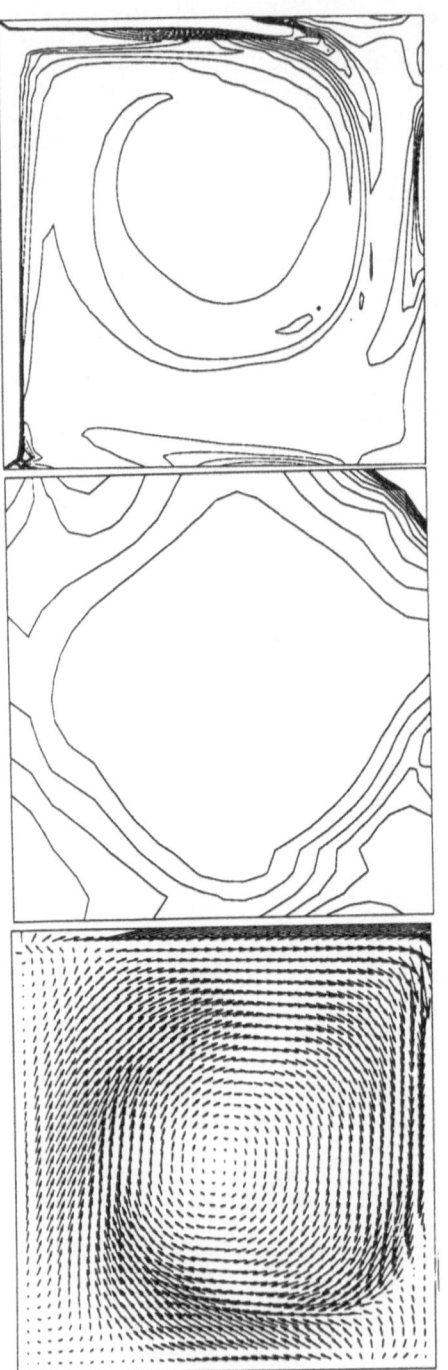

Contours of normal vorticity
at plane z=0
ω_{min} : -5 , ω_{max} : 5

T=200

Pressure iso-values at plane z=0
P_{min} : -0.005 , P_{max} : 0.0255

T=200

Velocity field at plane z=0
T=200

Figure 2

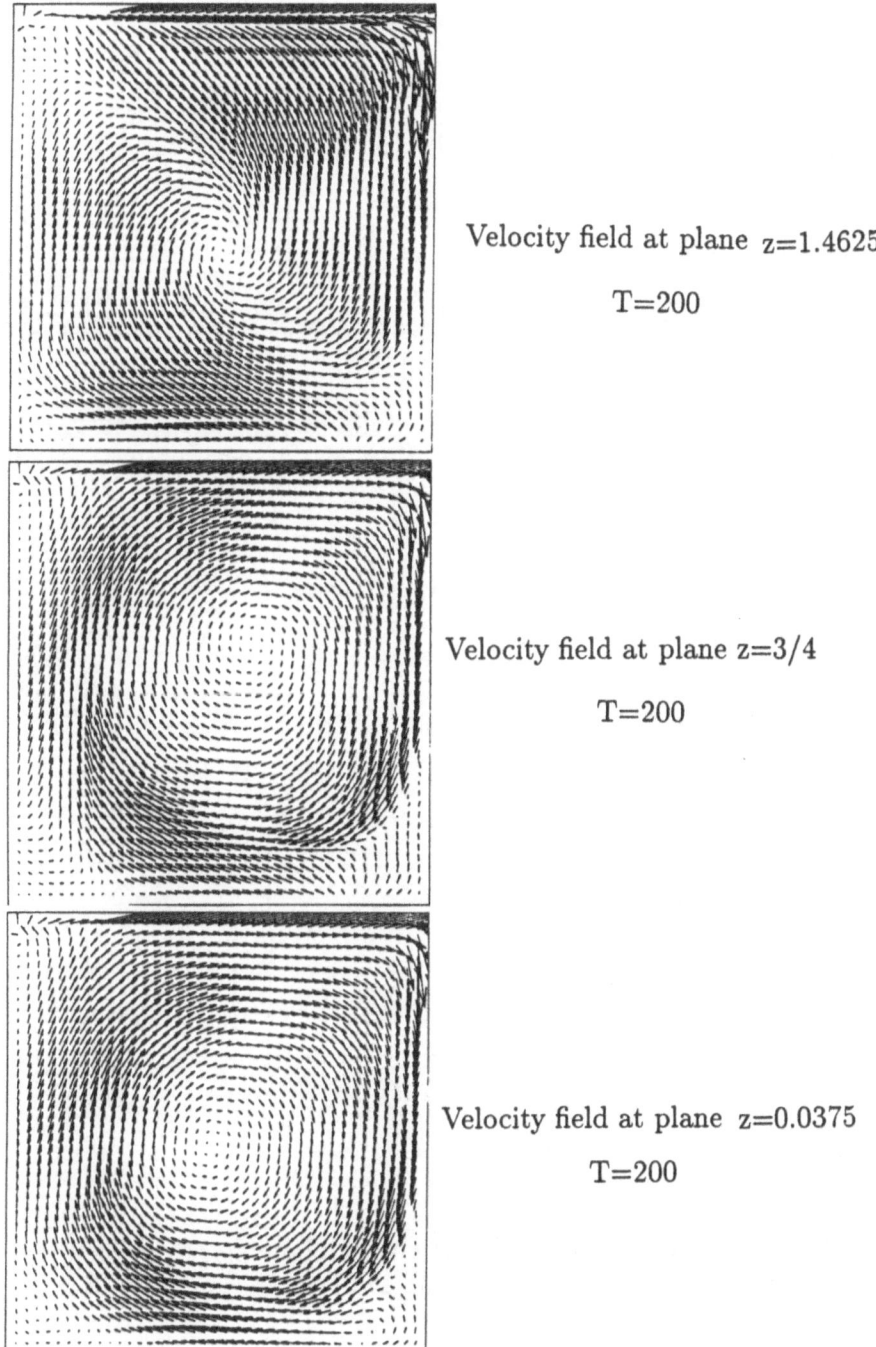

Velocity field at plane $z=1.4625$

T=200

Velocity field at plane $z=3/4$

T=200

Velocity field at plane $z=0.0375$

T=200

Figure 3

T=50

T=100

T=200

Contours of normal vorticity at plane x:4/15
ω_{min} : -5 , ω_{max} : 5

Figure 4

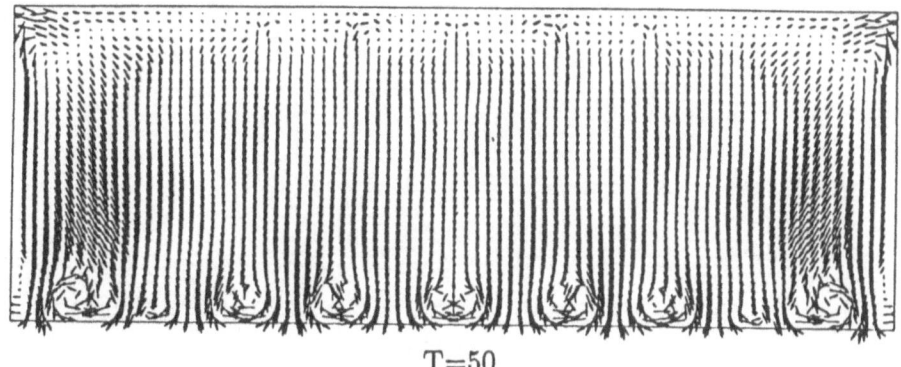

T=50

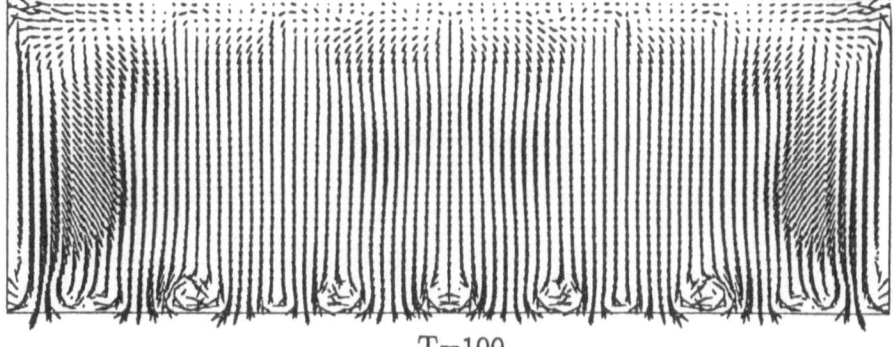

T=100

T=200

Velocity field at plane x :4/15

Figure 5

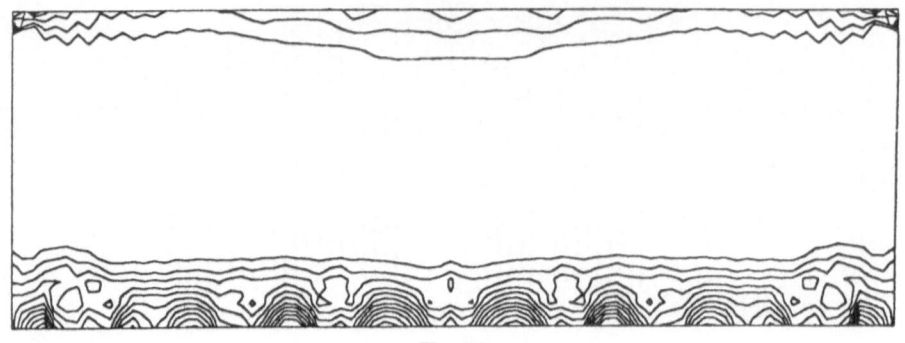

T=50

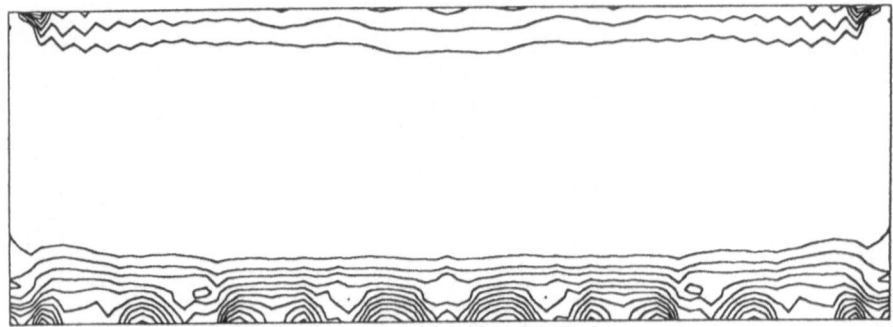

T=100

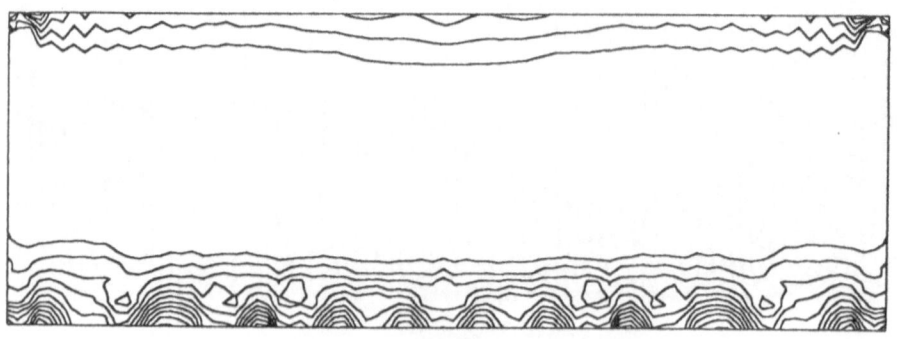

T=200

Pressure iso-values at plane x:4/15
P_{min} : -0.005 , P_{max} : 0.0255

Figure 6

Computation of 3-D unsteady laminar viscous flow
over a prolate spheroid at incidence
by a collocated finite difference method

PARIS GAMM WORKSHOP. JUNE 1991.

G.B. Deng, E. Guilmineau, P. Queutey & M. Visonneau.

CFD Group, LHN-URA1217 CNRS
ENSM, 1 rue de la Noë, 44072 NANTES Cedex, France.

Abstract

A general fully elliptic numerical method for the time dependent incompressible Navier-Stokes equations (DNS) has been developed and applied to the lifting flow past a prolate spheroid (Re=500, incidence of 30 degrees). The method retains some features of a general elliptic solver for the Reynolds-averaged-Navier-Stokes equations (RANSE) for which comparisons to experiments [1,2,3] and to others numerical solvers [4,5] have been done on the 6:1 prolate spheroid at high Reynolds numbers and various incidences [6,7].

The method uses a system of numerically generated curvilinear coordinates and retains the cartesian velocity components as dependent variables on a non-staggered grid.

1 THE NAVIER-STOKES EQUATIONS IN PRIMITIVE VARIABLES

1.1 Cartesian coordinates and cartesian components

The dimensionless Navier-Stokes equations applied to an incompressible fluid of dynamic viscosity ν are written in cartesian coordinates (x,y,z) for the primitive cartesian components U_k (k=1,2,3) of the velocity field U and the scalar pressure field p.

$$\frac{\partial U}{\partial t} + (U.\nabla).U + \nabla p = \frac{1}{Re}.\Delta U \tag{1.1}$$

$$DivU = 0. \tag{1.2}$$

The Reynolds number is defined as $Re = \dfrac{U_\infty L}{\nu}$. For an external flow, L is taken as a characteristic body length and U_∞ is taken as the free stream velocity modulus.

1.2 Partial transformation : generalized coordinates and cartesian components

In order to allow the study of the flow field past complex geometries, the curvilinear body-fitted coordinates (ξ,η,ζ) are introduced but the dependent variables are the cartesian velocity components and the scalar field p. This partial transformation of equations (1.1) and (1.2) yields the following equations,

$$J.\frac{\partial}{\partial t}(U_k) + \frac{\partial}{\partial \xi^i}(Ju^iU_k) + \frac{\partial}{\partial \xi^i}(b_k^ip) = \frac{\partial}{\partial \xi^i}\left(\frac{Jg^{ij}}{Re}.\frac{\partial U_k}{\partial \xi^j}\right). \tag{1.3}$$

$$\frac{\partial(Ju^i)}{\partial \xi^i} = 0. \tag{1.4}$$

The dimensionless mass fluxes Ju^i are defined by the projections,

$$Ju^i \equiv b^iU. \tag{1.5}$$

The Jacobian of the transformation from the curvilinear coordinates to the cartesian coordinates, $J=D(x,y,z)/D(\xi,\eta,\zeta)$, homogeneous to a volume, can be expressed by the three equivalent scalar products :

$$J = (x_\xi,y_\xi,z_\xi)^T.b^1 = (x_\eta,y_\eta,z_\eta)^T.b^2 = (x_\zeta,y_\zeta,z_\zeta)^T.b^3 \tag{1.6}$$

Elements $b^i = (b_1^i,b_2^i,b_3^i)^T$, the oriented areas in the surface ξ^i=const., are given by :

$$(b_j^i) = \begin{bmatrix} y_\eta z_\zeta - y_\zeta z_\eta & y_\zeta z_\xi - y_\xi z_\zeta & y_\xi z_\eta - y_\eta z_\xi \\ x_\zeta z_\eta - x_\eta z_\zeta & x_\xi z_\zeta - x_\zeta z_\xi & x_\eta z_\xi - x_\xi z_\eta \\ x_\eta y_\zeta - x_\zeta y_\eta & x_\zeta y_\xi - x_\xi y_\zeta & x_\xi y_\eta - x_\eta y_\xi \end{bmatrix} \tag{1.7}$$

Finally, the contravariant metric tensor, g^{ij}, is $g^{ij} = \dfrac{b^i.b^j}{J^2}$. \tag{1.8}

2 THE NUMERICS.

2.1 A generic form of the momentum equations

Starting from the partially transformed equations of conservation of momentum and assuming the orthogonality of the curvilinear coordinate system, equation (1.3) can be written as :

$$J.\frac{\partial U_k}{\partial t} + \frac{\partial I^i(U_k)}{\partial \xi^i} = 0 \tag{2.1}$$

where the momentum flux I^i of U_k is defined by (2.2)

$$I^i(U_k) = \frac{Jg^{ii}}{Re}.\frac{\partial U_k}{\partial \xi^i} - Ju^iU_k - b_k^i p \ . \tag{2.2}$$

2.2 Linearization of convection terms

The following linearization is used :

$$(Ju^iU_k)^n = (Ju^i)^{n-1}.U_k^{\ n} \ , \ n=1,2,... \tag{2.3}$$

For each time step t, $(Ju^i)^0$ is taken equal to the computed value at the previous time step.

2.3 A one dimensional exponential reconstruction scheme

2.3.1 A local integration

The one dimensional model for (2.1) is the conservation of a flux I, linear with respect to the dependant variable $\phi(x)$. I is assumed to vary locally on an interval [0,h], as (Figure 1),

$$I\big(\phi(x)\big) = I(x) = d.\frac{d\phi(x)}{dx} - c.\phi(x) - (\Sigma_0 + \sigma x) \ , \ \sigma = \frac{\Sigma_1 - \Sigma_0}{h} \ . \tag{2.4}$$

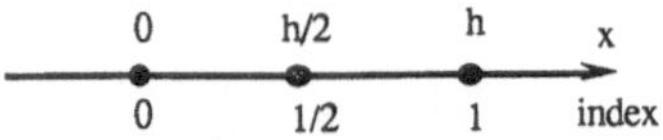

0	h/2	h	x
0	1/2	1	index

Figure 1

● Definition nodes for ϕ and Σ
◉ Definition nodes for c and d

The known linearized source term $\Sigma(x) = (\Sigma_0 + \sigma x)$ mimics the pressure field while c and d stand for the convection and the diffusion coefficients which are taken as local constants .

To proceed further, we suppose that I(x) is locally constant over [0,h] and takes the unknown value I. Then, a simple quadrature from 0 to x gives the locally analytic expression of the dependent variable,*shape function* :

$$\phi(x) = e^{Bx}\phi_0 - \frac{1}{c}\left(\sigma.\left(x + \frac{1}{B} - \frac{e^{Bx}}{B} \right) + (I+\Sigma_0).(1 - e^{Bx}) \right) \ , \ B \equiv \frac{c}{d}. \tag{2.5}$$

The second boundary condition at x=h, $\phi(h) = \phi_1$, fixes the flux I (γ is the local mesh Peclet number) :

$$I = \frac{c}{e^\gamma - 1}.\phi_1 - \frac{ce^\gamma}{e^\gamma - 1}.\phi_0 - \sigma h.\left(\frac{e^\gamma - \gamma - 1}{\gamma.(e^\gamma - 1)}\right) - \Sigma_0 \ , \ \gamma \equiv \frac{ch}{d} \ . \tag{2.6}$$

2.3.2 The one-dimensional scheme

From (2.5) and (2.6), the dependent variable ϕ can be reconstructed at the mid-point x=h/2 :

$$\phi_{1/2} \equiv \phi(h/2) = \beta_0.\phi_0 + \beta_1.\phi_1 - \beta_\sigma.\sigma h \ ,$$

$$\beta_0(\gamma) = \frac{e^{\gamma/2}}{e^{\gamma/2}+1} \quad , \quad \beta_1(\gamma) = 1 - \beta_0(\gamma) = \frac{1}{e^{\gamma/2}+1} \quad , \quad \beta_\sigma(\gamma) = \frac{h}{8d}.\frac{Tanh(\gamma/4)}{\gamma/4} \ . \tag{2.7}$$

The one-dimensional scheme is obtained by considering two adjacent cells of respective lengths h_m and h_p :

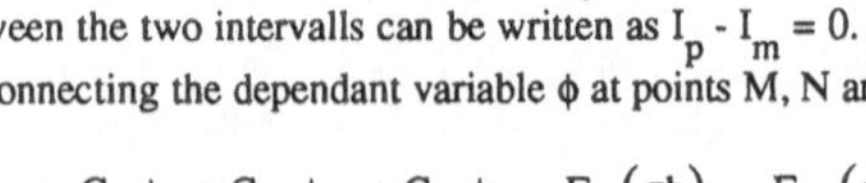

-h_m		0		+h_p	
M	m	N	p	P	x

Figure 2

The conservation of I between the two intervalls can be written as $I_p - I_m = 0$. Using (2.6) this flux balance gives a compact relation connecting the dependant variable ϕ at points M, N and P to the source gradient at mid-points m and p :

$$I_p - I_m = C_P.\phi_P + C_M.\phi_M + C_N.\phi_N - \Gamma_p.(\sigma h)_p - \Gamma_m.(\sigma h)_m \ . \tag{2.8}$$

With the the definition of coefficients,

120

$$C_P = \frac{c_p}{e^{\gamma_p} - 1} \ , \ C_M = \frac{c_m e^{\gamma_m}}{e^{\gamma_m} - 1} \ , C_N = \left(\frac{c_p e^{\gamma_p}}{e^{\gamma_p} - 1} + \frac{c_m}{e^{\gamma_m} - 1} \right) = - C_P - C_M - [\, c_p - c_m \,] \ , \quad (2.9)$$

$$\Gamma_p = \left(\frac{e^{\gamma_p} - \gamma_p - 1}{\gamma_p.(e^{\gamma_p} - 1)} \right) \ , \ \Gamma_m = \left(\frac{e^{-\gamma_m} + \gamma_m - 1}{\gamma_m.(1 - e^{-\gamma_m})} \right).$$

Du to upwinding of the source gradient, this scheme is equivalent to a centered scheme, the location of the central point depending on the mesh Peclet number. The accuracy and the stability of the scheme are not restricted to limited values of the mesh Peclet number. One can notice that a *collocation* between the variables (φ and Σ) is allowed because of the locally constant hypothesis over the gradient term .

2.4 Multidimensional extension and application to the 3DNS equations

We now suppose that the fluxes are constant over prescribed control volumes of the computational domain (Figure 3). A typical cell is limited by the mid-points (m_i, p_i) in each direction. An integration of the conservation equation (2.1) yields :

$$\left(J.\frac{\partial U_k}{\partial t} \right)_N + \frac{I^i_{p_i}(U_k) - I^i_{m_i}(U_k)}{\Delta \xi^i} = 0 \qquad (2.10)$$

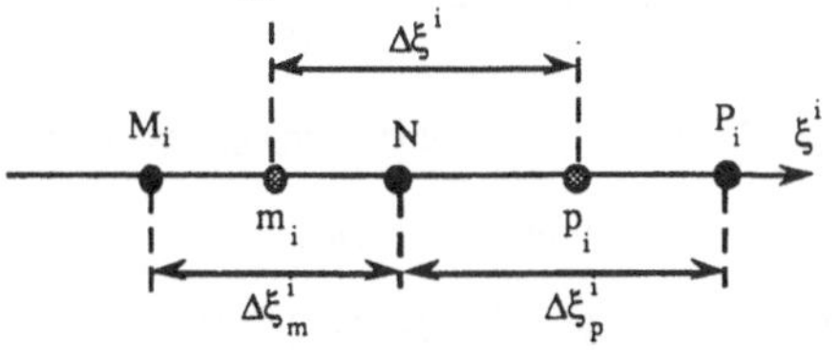

Figure 3

The mid-point fluxes are reconstructed using (2.6) and the discrete relation involving neighbours of N is,

$$\left(J.\frac{\partial U_k}{\partial t} \right)_N + C_N.U_{kN} + \sum_i \left(C_{P_i}.U_{kP_i} + C_{M_i}.U_{kM_i} \right) = \sum_i \left(\Gamma_{p_i}(\sigma^i_k)_{p_i} + \Gamma_{m_i}(\sigma^i_k)_{m_i} \right). \quad (2.11)$$

The terms σ^i_k involve discrete pressure gradients at mid-points m_i and p_i :

$$(\sigma^i_k)_{p_i} = (b^i_k)_{p_i}.(P_{P_i} - P_N) \ , \quad (\sigma^i_k)_{m_i} = (b^i_k)_{m_i}.(P_N - P_{M_i}) . \qquad (2.12)$$

Now, the mesh Peclet numbers to be considered are defined by :

$$\gamma_{s_i} = \frac{c^i_s \Delta \xi^i_s}{d^i_s} \quad \text{with} \quad c^i_s \equiv (Ju)^i_{s_i} \quad \text{and} \quad d^i_s \equiv \left(\frac{Jg^{ii}}{Re} \right)_{s_i} \ , \text{ for } s = m \ \& \ p. \qquad (2.13)$$

For simplicity let us introduce the discrete "div" operator applied to the vector A of contravariant components A^i,

$$\text{div}(A) = \sum_i \left(\frac{A^i_{p_i} - A^i_{m_i}}{\Delta \xi^i} \right) . \qquad (2.14)$$

Then, the C and Γ coefficients are deduced from (2.9) to give (2.15)

$$C_{P_i} = \frac{1}{\Delta \xi^i_p} \left(\frac{b^i_p}{e^{\gamma_{p_i}} - 1} \right) \ , C_{M_i} = \frac{1}{\Delta \xi^i_m} \left(\frac{b^i_m e^{\gamma_{m_i}}}{e^{\gamma_{m_i}} - 1} \right) \ ,$$

$$\Gamma_{P_i} = \frac{1}{\Delta \xi^i_p} \left(\frac{e^{\gamma_{p_i}} - \gamma_{p_i} - 1}{\gamma_{p_i}.(e^{\gamma_{p_i}} - 1)} \right) \ , \Gamma_{m_i} = \frac{1}{\Delta \xi^i_m} \left(\frac{e^{-\gamma_{m_i}} + \gamma_{m_i} - 1}{\gamma_{m_i}.(1 - e^{-\gamma_{m_i}})} \right) \ ,$$

$$C_N = - \sum_i \left(C_{P_i} + C_{M_i} \right) - \text{div}(Ju) . \tag{2.15}$$

It should be noticed that $\text{div}(Ju)$ in (2.15) is a discrete approximation of $J.\text{Div}(U)$ and vanishes when the non-linearities are solved, if the continuity constraint is discretized as $\text{div}(Ju)=0$.

2.5 Treatment of the continuity constraint

The continuity constraint is discretized in the same way as the momentum balance studied previously. The mass-fluxes b^i_s introduced in (2.12) are reconstructed according to the one dimensional scheme (2.7) and provide at point $s=m_i$:

$$(Ju^i)_{m_i} = (b^i.U)_{m_i} = (Ju^{*i})_{m_i} - (b^ib^i)_{m_i} \beta_{m_i}(\gamma_{m_i}).[\, p_N - p_{M_i}\,] ,$$

$$(Ju^{*i})_{m_i} = \beta_{M_i}(\gamma_{m_i}).(b^i)_{m_i}(U)_{M_i} + \beta_N(\gamma_{m_i}).(b^i)_{m_i}(U)_N . \tag{2.16}$$

The reconstruction coefficients are issued from (2.10) if $\beta_{M_i}=\beta_0$, $\beta_N=\beta_1$ and $\beta_{m_i}=\beta_\sigma$. Then, from the reconstruction (2.16), the continuity constraint can now be written as (2.17) using the contravariant pseudo-vector $Ju^* = (Ju^{*i})$,

$$- \text{div}(Ju^*)_N = \sum_i \left(\frac{(b^ib^i)_{p_i}.\beta_{p_i}(\gamma_{p_i}).[\, p_{P_i} - p_N\,] - (b^ib^i)_{m_i}.\beta_{m_i}(\gamma_{m_i}).[\, p_N - p_{M_i}\,]}{\Delta\xi^i} \right) \tag{2.17}$$

Relation (2.17) can be interpreted as a pressure equation emerging from the discretisation of the continuous operator :

$$L(p) = - \text{Div}(Ju^*) \tag{2.18}$$

Du to the positive nature of the β coefficient, (2.18) generates linear system with a *symmetric and positive definite banded matrix*. In this way, the elliptic nature of the pressure is preserved.

2.6 Treatment of the unsteady term
2.6.1 Approximation of the time derivative

The time problem being parabolic, the generic term $\frac{\partial\phi}{\partial t}$ is backward differenced. We are considering the time derivative of ϕ at point N for time t where $\phi=\phi^1=\phi_N$ is unknown. We notice that ϕ at previous times, namely ϕ^0, ϕ^{00} ..., are known (see Figure 4). The time derivative at t can be approximated by :

$$\frac{\partial\phi}{\partial t} = \frac{\phi^{1/2} - \phi^{-1/2}}{(\Delta t_0 + \Delta t_1)/2} \tag{2.19}$$

Using Taylor series into past around the mid-points 1/2 and -1/2, we have the time reconstructions :

$$\phi^{1/2} = \phi^0 + (\phi^1 - \phi^0).\left(\frac{\Delta t_0 + \Delta t_1/2}{\Delta t_0}\right) \quad , \quad \phi^{-1/2} = \phi^{00} + (\phi^0 - \phi^{00}).\left(\frac{\Delta t_{00} + \Delta t_0/2}{\Delta t_{00}}\right). \tag{2.20}$$

Substituing (2.20) into (2.19) gives :

$$\frac{\partial\phi}{\partial t} = \frac{\alpha\phi^1 - \phi^{0*}}{\Delta t_e} \quad , \quad \Delta t_e = (\Delta t_0 + \Delta t_1)/2 \ ,$$

with
$$\begin{array}{|l}
\text{first order} \quad : \alpha = 1 \qquad \phi^{0*} = \phi^0 , \\[2em]
\text{second order} \quad : \alpha = 1 + \frac{\Delta t_1}{2\Delta t_0} \quad \phi^{0*} = \phi^0.\left(1 + \frac{\Delta t_1}{2\Delta t_0} + \frac{\Delta t_0}{2\Delta t_{00}}\right) - \phi^{00}.\left(\frac{\Delta t_0}{2\Delta t_{00}}\right).
\end{array} \tag{2.21}$$

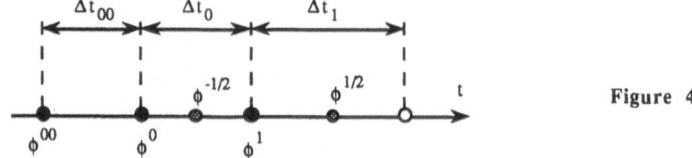

Figure 4

2.6.2 Definition of the variable time step : $\Delta t(t)$

Because unsteady simulations are cpu-time consuming, it has been found necessary to minimize the number of time steps. Starting from an initial known field at t_i, the final time t_f is reached in n steps using an increasing time step $\Delta t(t)$ defined by :

$$\Delta t(t) = (\Delta t_i - \Delta t_f).Exp\left(- \frac{t - t_i}{\tau} \right) + \Delta t_f . \qquad (2.22)$$

It is necessary to specify τ, the characteristic time ; Δt_i, the initial time step at t_i ; Δt_f, the final time step (when $t \to +\infty$). The influence of the characteristic time τ on $\Delta t(t)$ is indicated in fig.5.

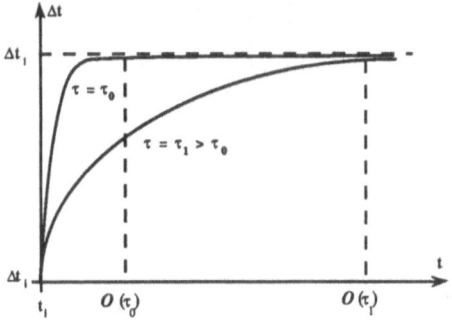

Figure 5

It can be noticed that the variation of the time step at t_i is exactly given by $\left(\frac{\partial \Delta t}{\partial t}\right)_{t=t_i} = \frac{\Delta t_f - \Delta t_i}{\tau}$.

2.6.3 Unsteady numerical scheme

Since the generic time derivative is defined, we obtain the final numerical scheme for the momentum equations. The relation between points around N at time t is, from (2.14) ,

$$\left(\frac{\alpha J_N}{\Delta t_e} + C_N\right)\phi_{kN} + \sum_i \left(C_{P_i}.\phi_{kP_i} + C_{M_i}.\phi_{kM_i} \right)$$

$$= \sum_i \left(\Gamma_{pi}(\sigma_k^i)_{p_i} + \Gamma_{m_i}(\sigma_k^i)_{m_i} \right) + \left(\frac{J_N}{\Delta t_e}\right)\phi_k^{0*} . \qquad (2.23)$$

3 SOLUTION STRATEGY AND NUMERICAL SOLVER(S)

3.1 Choice of dependent variables layout

The retained choice is the so-called "Collocated Cell-Centered" layout where the velocity field, through its cartesian components, and the pressure field are located at the center of the control volume defined by eight points of the structured mesh containing $N_1 \times N_2 \times N_3$ points, $\{x(i,j,k),y(i,j,k),z(i,j,k)\}$.

An inner view of the control volume in the computational domain with the position of unknowns and fluxes is presented in Figure 6. The three presented faces show some of the boundaries of the cell used to balance the momentum fluxes and the mass fluxes.

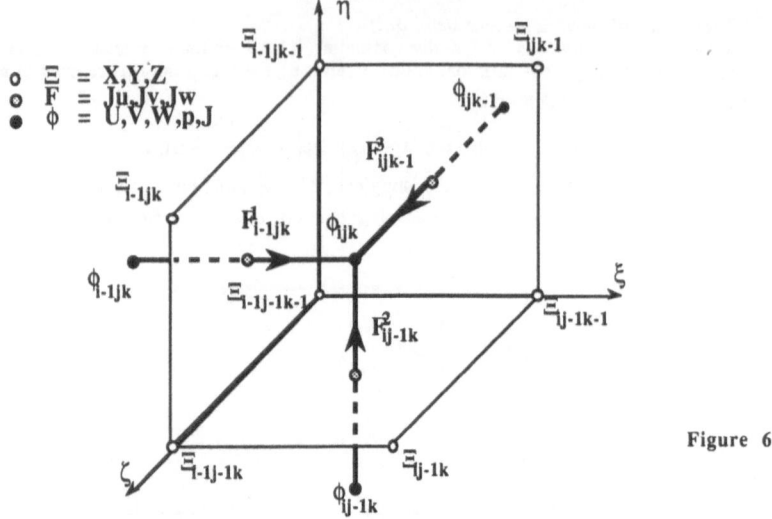

$$\begin{aligned} \circ \circ \quad \Xi &= X,Y,Z \\ \circ \circ \quad F &= Ju,Jv,Jw \\ \bullet \quad \phi &= U,V,W,p,J \end{aligned}$$

Figure 6

On the boundary surfaces, the dependent variables (U,V,W,p) are located at the center of the elementary boundary surfaces generated by the four grid points (o). This cell-centered choice simplifies the treatment of the continuity constraint and the imposition of boundary conditions.

3.2 Boundary conditions

In order to simplify the argument, only the case of a given boundary velocity field U_b is considered. Then, if $\partial\Omega$ is the closed bounded surface of the computational domain Ω and $\mathbf{n}$ is the outward normal, U_b must verify the integral identity :

$$\int_{\partial\Omega} U_b \cdot \mathbf{n} = 0 . \qquad (3.1)$$

This is the choice here where U_b is taken equal to U_∞ far from the prolate spheroid. In this last case the boundary mass fluxes are given without mass loss. The continuity constraint (2.20), with a divergence free velocity vector, implies that the elliptic pressure problem provides an infinity of solutions within a constant (if p is solution then p+constant must be also solution). Considering a one dimensional model problem, figure 7 shows the n-1 elementary control volumes for continuity constraint for n discretisation points,

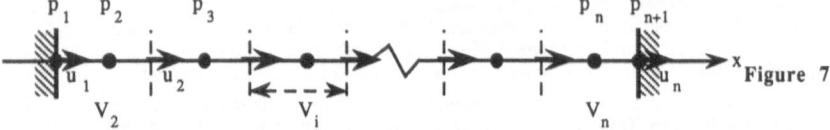

Figure 7

Relation (2.20) applied to the elementary contol volume V_i, namely $u_i - u_{i-1} = 0$, yields a relation between the two pressure gradients $(p_{i+1} - p_i)$ and $(p_i - p_{i-1})$ with the difference pseudo-mass fluxes $(u_i^* - u_{i-1}^*)$. On the left boundary, where mass flux u_1 is known (no reconstruction needed for u_1!), conservation of mass through V_2 involves only $(p_3 - p_2)$ as function of the difference mass flux $(u_2^* - u_1)$. So the problem is closed for the unknown gradients $(p_{i+1} - p_i)$, i=2,n-1 and defines the pressure field within an arbirary constant. Because pressures (or pressure gradients) are needed at boundaries for the momentum equations, the boundary values are extrapolated at first order for the gradient (second order for the pressure) :

$$\left(\frac{\partial p}{\partial x}\right)_1 = \left(\frac{\partial p}{\partial x}\right)_2, \text{ approximated by } p_1 = \frac{3p_2 - p_3}{2}, \text{ for constant size control volumes.} \qquad (3.2)$$

124

3.3 Solution(s) algorithm(s)

Equations to be solved, (2.22) and (2.28) with appropriate boundary conditions yield, for the set of standard points, the symbolic linear system :

$$\sum_{nb} C_{nb} \cdot X_{nb} = S_{nb} \,.$$ (3.3)

The index nb stands for $\{N,M_i,P_i\}$, X is the vector of unknowns $(U,V,W,p)^T$ and S is the explicit source term containing Dirichlet boundary conditions and the solution at the previous time step. Each elementary bloc coefficient matrix C_{nb} has the following structure, entirely dictated by the linearisation method and the reconstruction scheme for fluxes,

a1	0	0	b1
0	a2	0	b2
0	0	a3	b3
c1	c2	c3	a4

(3.4)

Coefficients ai and bi (for i=1,2 and 3) are the coefficients of the linearised discretised momentum equations (2.22) applied respectively to U_k (U,V and W) and to p on the current control volume. Coefficients ci (for i=1,2, and 3) and a4 have the same signification but when applied to the discretized continuity constraint (2.20). Coefficients bi and ci (for i=1,2 and 3) are *coupling coefficients*. When ci and bi are treated as source terms, a *decoupled algorithm* is generated at each time step consisting in,

   (D1) solution of the three momentum equations to get U,V and W ;
   (D2) solution of *the continuity equation seen as a pressure equation* ;
   (D3) reconstruction of the convective fluxes (uptading of convective fluxes) ;
   (D4) updating the solution for non-linearities.

Such an algorithm is similar in nature to the classical PISO type algorithms, depending on a possible correction step between (D2) and (D4) in order to re-enforce the coupling between the velocity field and the pressure field.

The *coupled algorithm* has the following elementary steps consisting in ,

   (C1) solution of the coupled linear system to get U,V,W and p ;
   (C2) reconstruction of the convective fluxes (same operation as (D3));
   (C3) updating step for the solution for non-linearities (same operation as (D4)).

For instance our choice is to use a coupled algorithm for 2D problems using a preconditionned biconjugate gradient method or a conjugate gradient squared method. For 3D problems, a decoupled algorithm with a point-Jacobi method is used for solving the momentum equations and a symmetrically preconditionned conjugate gradient method to solve the "pressure equation". All these solvers have been written in order to provide a high flexibility for several types of boundary conditions, also they are optimized so as to reach significant speed-up ratios on vector computers.

4 VALIDATION OF THE METHOD A 2D MODEL PROBLEM

This model problem is the impulsively started circular cylinder at Re=20 for which comparisons with experiments [8] are possible. This problem is summarized at Figure 8. The reference length is D, the diameter of the cylinder (R=D/2), and the reference velocity is U_∞ , the free stream velocity modulus. The mesh size is $(N_1 x N_2)$ and for boundary conditions, the line (y=0,x>0) is the line of periodicity for the curvilinear body fitted lines ξ, i=1 and i=N_1; the no-slip at j=1 and U=U_∞,V=0 at line j=N_2.

With a (65x65) size mesh, an exponential stretching is used into the radial direction such that distance between lines j=2 and j=1 is 0.03R and the far-field line j=N_2 is positionned at 50 radii from the origin (0,0). The pressure reference point is located into the upstream wake on y=0.

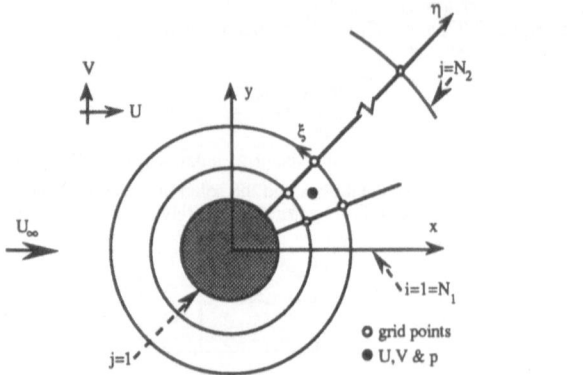

Figure 8

Starting with (U,V)=(1,0) everywhere at time t=0, the computations proceede at t>0 with the previously prescribed boundary conditions and the first order time scheme. For the Reynolds number Re=20, a stationnary flow field is found at about t=7.0. Results are compared to experimental values [8] of U on the centerline y=0 and x>0 , at controlled times 2.3, 4.0, and 10.8. The time step law is built with Δt_i =0.1, Δt_f =0.5 and τ=10.0. **Figure 9** compares the centerline velocity to experiments at control times.

Computations have been performed on a VP200 on which the vectorisation rate for the 2D-coupled code is of 96.3%. At each time step, non-linearities are solved until the maximum non-linear residual is reduced by 6 orders. The mean cost for computing one time step is about 2.5 seconds (for the (65x65) mesh) or 0.61 msec. per point or 1.4 10^{-5} sec. per point per iteration (iterations for non-linearity).

5 APPLICATION TO THE 6:1 PROLATE SPHEROID AT INCIDENCE

5.1 Sketch definition

The half plane z>0 is considered an as showned at figures 10 and 11, the mesh used is of O-O type with two singular lines, the polar axii starting from the nose and the aft point, where the Jacobian vanishes.

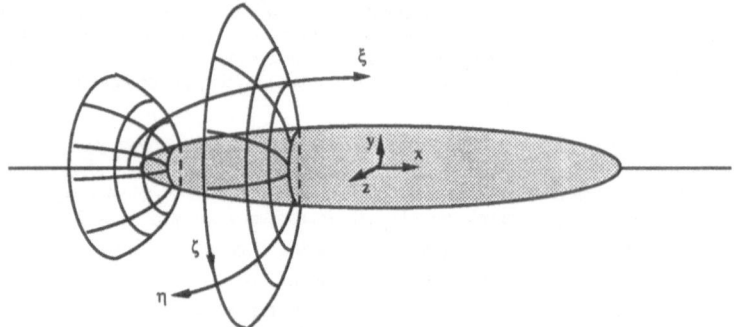

Figure 10

Referring to (i,j,k) indexes for body-fitted (ξ,η,ζ) coordinates, these singular lines are respectively i=1 and i=N1. The body surface is j=1 where no-slip condition is imposed and j=N2 is the far field surface where the velocity is set equal to the free stream velocity : $U_\infty=\cos(\alpha)$,$V_\infty=\sin(\alpha)$,$W_\infty=0$, if α is the incidence of the flow. In the symmetry planes, at z=0, symmetries (on U,V or p) or anti-symmetry (on W) are written by adding symmetric extra planes in z<0. This planes are useful for the computation of metrics at z=0. On the singular lines the transformation of coordinates (see 1.1) is no longer valid, but, because of multiple definition of points (1,j,k) and (N1,j,k), the mass fluxes into direction ξ are zero through the vanishing areas i=1 and i=N1.

The mesh is built with 61 longitudinal points (ξ), 50 normal points (η) and 41 circumferential points (ζ). The external surface is located at 10L, if L is the lenght of the major axis of the spheroid. The sinh stretching used to cluster the points near the solid surface is such that the distance between surfaces j=2 and j=1 is about 0.001L. Points are also clustered near the nose and the aft point and close to the leeside symmetry plane where circumferential separation is expected, due to the vortical nature of the flow.

126

5.2 Results

For this Workshop special attention was required at times t=1,2,3,4 and 5. A constant time step of 0.1 is used for this problem and at time t=0 the flow field is at rest. As shown through the time variation of the pressure and the z-component (Wz) of the vorticity field (W) at the leading edge (-0.5,0,0), a steady state is reach at about t=4. Only a first order time scheme has been used for these computations. For the treatment of the numerical results, a three dimensional linear interpolation was used to plot most of the required figures in x, y or z planes.

The strong wall pressure gradients around the stagnation point (in the windside plane) seem to be well captured with a sufficient stretching (the squares indicate the location of the discretization points).

Similarly, the wall z-component, $Wz=\partial V/\partial x-\partial U/\partial z$, of the vorticity field has extrema values near the leading edge and the trailing edge due to acceleration of the flow field to pass around the body. Circumferentially, from the windside plane, strong shear stresses are detected, through the plotting of 11 levels of iso-Wx. The crossflow separation occuring at the leeside is not directly correlated with the shape of iso-pressure lines. No leeside wall pressure plateau was detected after the crossflow separation line, on the contrary to higher Reynolds numbers. Also pointed out, the fact that no secondary separation occurs contrary to [9] for Re=10,000. **Figure 12** shows the skin-friction lines on the body surface and the plne z=0 colored by the pressure field (high levels in red, low levels in blue). **Figure 13** presents the secondary velocities at plane x=0 at the leeside where crossflow separation is always developed. This last picture is colored by the magnitude of the x-component of the vorticity field.

Computations where performed on a VP200. A 96% vectorization rate was reached for the 3D solver written in Fortran-77$^{©}$. The computational cost for solving the four decoupled linear systems (U,V,W and p) is 0.045 msec per point (a residual reduction of 8 orders is required for the momentum equations and of 3 orders for the 'pressure equation'). At each time step the non-linear residuals are decreased by 2.5 to 3 orders with 150 non-linear iterations. Then, for the 61x50x41 mesh, the time to compute one time step is about 8 minutes.

Acknowledgements. Thanks are due to the Scientific Commitee of CCVR and the DS/SPI for attributions of Cpu on the VP200. Flow visualizations in figures. 11, 12 and 13 have been performed with the software 'ASCETE' from the CFD Group.

References

[1] **Meier,H.U. & Kreplin,H.P.** "Experimental investigation of the boundary layer transition and separation on a body of revolution", Z. Flugwiss. Weltraumforsch., 4, Heft 2, 65-71 (1980).

[2] **Meier,H.U., Kreplin,H.P.,& Vollmers,H.** "Development of boundary layers and separation patterns on a body of revolution at incidence." Proc. 2nd. Symp. on Num. and Phys. Aspects of Aerodynamic Flows, Long Beach, Cal. (1983).

[3] **Coponet,D. & Soulevant,D.,**"Etude par velocimétrie laser de la couche limite sur un ellipsoïde de révolution allongé du DFVLR"; Annexe n°1 du PV 4/1752 AN Internal Report ONERA (1984).

[4] **Vatsa,V.N., Thomas,J.L. & Wedan, B.W.** "Navier-Stokes computation of prolate spheroids at angle of attack". J. Aircraft, 26, 986-993 (1989). See also AIAA 87-2627-CP (1987).

[5] **Wong,T.C., Kandil,O.A. & Liu, C.H.** "Navier-Stokes computation of separated vortical flows past a prolate spheroid at incidence", AIAA 89-0553 (1989).

[6] **Piquet,J. & Queutey, P.** "Computation of the viscous flow past a prolate spheroid", Proc. 8th GAMM Conf. Num. Meth. Fluid Dynamics, Wesseling, P.Ed., Vieweg (1989).

[7] **Deng,G.B., Piquet,J. & Queutey,P.** "Navier-Stokes computation of vortical flows", AIAA 89-1625 (1989).

[8] **Bouard,R. & Coutanceau,M.** "Experimental determination of the main feature of the viscous flow in the wake of a circular cylinder in uniform translation". Journal of Fluid Mech., Vol 79, p. 231 (1977).

[9] **Costis,C.E., Polen,D.M. , Hoang,N.T. & Telionis,D.P.,** "Laminar separating flow over a prolate spheroid". AIAA 87-1212, (1987).

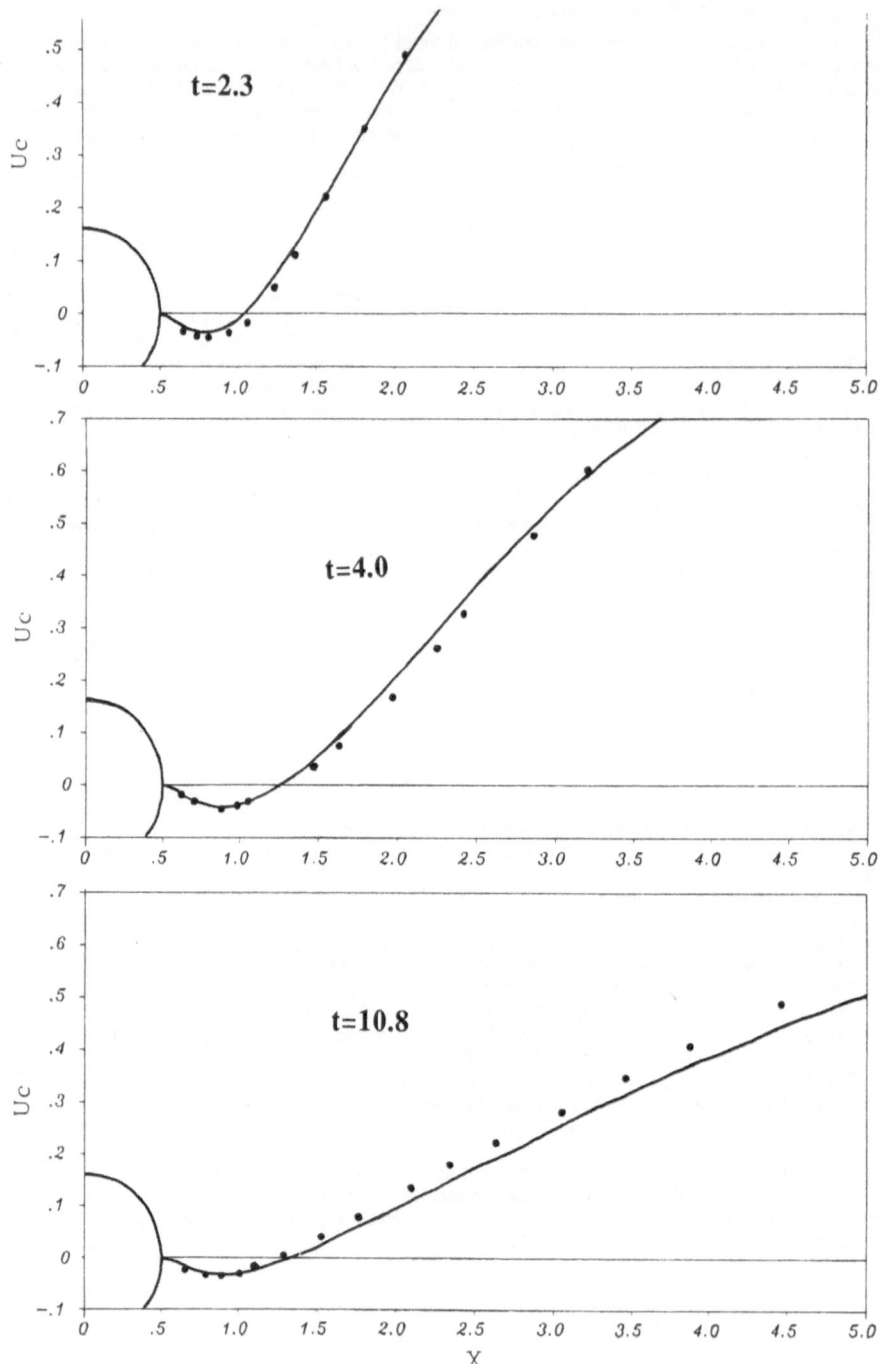

Figure 9 Velocity distribution on the wake center line.
(---- = computed   ● = experiments)

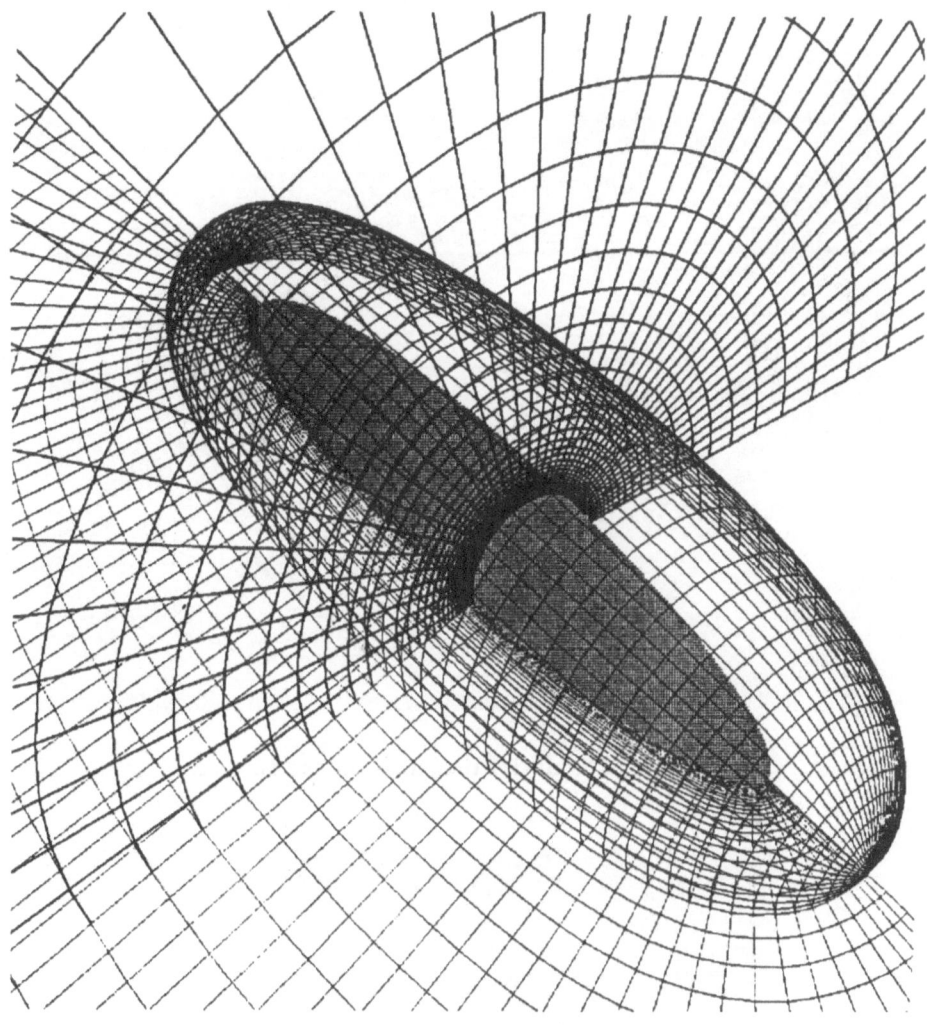

Figure 11 View of the grid $(\xi-61*\eta-50*\zeta-41)$.

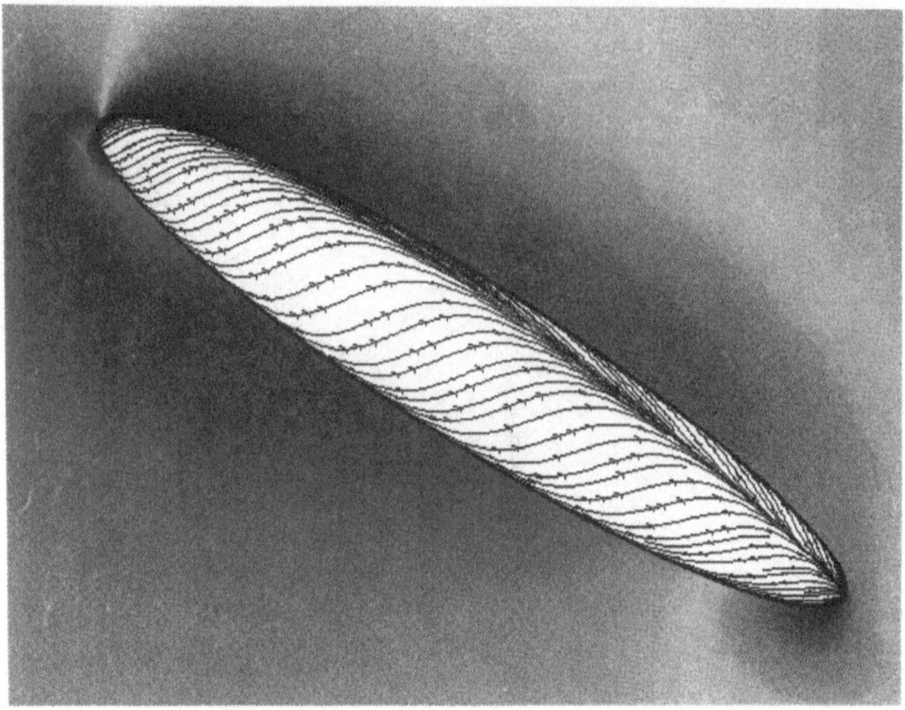

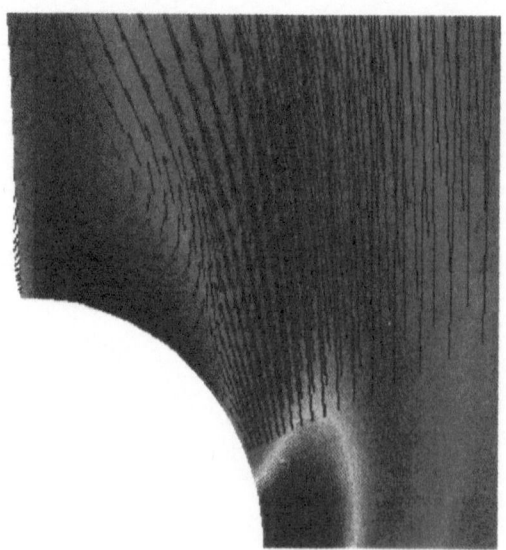

Figure 12 Skin friction lines on the surface.

ASCETE.2.0

Figure 13 Secondary flow at X/L=0.

Multidomain Technique for 3-D Incompressible
Unsteady Viscous Laminar Flow
around Prolate Spheroid

K.-C. Le Thanh

ONERA, BP 72, Châtillon, France.

Abstract : Navier-Stokes equations, written in velocity-pressure formulation are solved using finite-difference scheme in time and space. Velocity and pressure are discretized on a non-staggered grid. A residual smoothing technique is chosen allowing an increase of the domain of stability. We use a non-overlapping multidomain technique based on the computation of residuals at interfaces and at each step of the algorithm.

Introduction

To solve the 3-D Navier-Stokes equations for unsteady, incompressible viscous flow, we use a non-overlapping multidomain technique.

The original algorithm in cartesian coordinates was first developed for studies related to direct simulation of turbulence and vortex breakdown simulation [1,2]. The 3-D solver has been more recently developed and tested on an impulsively started spheroid [3,4].

Equations with Dirichlet boundary conditions are discretized on a non-staggered grid using a finite-difference scheme in space and time.

For time discretization, we use an Adams-Bashforth scheme for convective terms, and an explicit Euler scheme for diffusive terms. An implicit operator is introduced to enhance the stability of the scheme and to damp non linear oscillations [5].

By taking the discrete divergence of the discrete momentum equation, and expressing condition of divergence-free velocity field, we obtain a discrete equation to determine the total pressure. This equation is solved using a multigradient like method which minimizes a l^2-norm of the residual.

The subdomain technique is characterized by writing derivatives and residuals at each time step of the algorithm.

Analysis

We consider the Navier-Stokes equations for a viscous incompressible fluid in a bounded domain $\Omega \subset R^3$ with boundary $\Gamma = \Gamma^1 \cup \Gamma^2 = \partial \Omega$ where Γ^1 is the surface of the prolate spheroid, and Γ^2 is the boundary at infinity:

$$\begin{cases} \dfrac{\partial u}{\partial t} + (\nabla \times \underline{u}) \times \underline{u} + \nabla q - \nu \nabla \cdot (\nabla \underline{u}) = 0 , & in\ \Omega \quad (1) \\[2mm] \nabla \cdot \underline{u} = 0 , & in\ \Omega \quad (2) \end{cases}$$

$\underline{u} = (u,v,w)$ velocity expressed in cartesian components and $q = \frac{1}{2}\underline{u}^2 + p$ total pressure denote the unknowns, v is the kinematic viscosity, ∇ (respectively $\nabla \cdot$) is the gradient operator (respectively divergence operator).

The non-linear term is written in rotational form for reasons of computational efficiency [2].

To complete equations (1) and (2), the boundary and initial conditions of the physical problem are those required:

      - no slip on the surface of the prolate spheroid,
      - uniform flow at infinity,
      - initially, the body is at rest in the fluid,

these conditions can be written as:

    - $\underline{u}|_{\Gamma^1} = 0$
    - $\underline{u}|_{\Gamma^2} = u_\infty(t)$
    - $\underline{u}(\underline{x}, 0) = 0$.

The fonction $u_\infty(t)$ regular over $\overline{\Omega}$ satisfies : $u_\infty(0) = 0$ and $\lim\limits_{t \to +\infty} u_\infty(t) = u_\infty$ where u_∞ is the upstream uniform velocity ($u_\infty = (cos\,\alpha\,,sin\alpha\,,0\,)$ with $\alpha = 30°$).

Numerical method

Grid generation

The surface grid is obtained by projecting the grid of a cube surface onto the prolate spheroid creating thus a non-singular mesh. The whole grid is generated from the surface by regular transformation. Six subdomains are obtained, corresponding to the six faces of the cube. The mesh is conforming in the sense that mesh lines are continuous across interfaces between subdomains.

Grid points are concentrated near the body surface according to the Reynolds number for an adequate resolution of the boundary layer. The thickness of the boundary layer is estimated from the flat plate theory ($\delta \sim \frac{1}{\sqrt{Re}}$).

The boundary at infinity is the surface of a sphere of radius $3.5L$ where L is the prolate spheroid length.

Multidomain technique

The physical domain Ω is overlayed by six open subdomains Ω^l ($l \in [1,6]$) such that $\overline{\Omega} = \bigcup\limits_{l=1}^{6} \overline{\Omega}^l$ and domains Ω^l do not intersect each other.

We denote $\gamma^{m,n}$ the interface between two adjacent subdomains Ω^m and Ω^n ($\gamma^{m,n} = \overline{\Omega}^m \cap \overline{\Omega}^n$ ($m \neq n$)).

We can distinguish here two kinds of interfaces:

    - $\gamma^{m,n}$ is a surface,
    - $\partial\gamma^{m,n}$ is an edge and, there exists a subdomain $\Omega^p \subset \Omega$
      such that $\overline{\Omega}^m \cap \overline{\Omega}^n \cap \overline{\Omega}^p = \partial\gamma^{m,n} \cap \overline{\Omega}^p = \partial\gamma^{m,p} \cap \overline{\Omega}^n = \partial\gamma^{n,p} \cap \overline{\Omega}^m$.

The treated geometry requires the use of a generalized coordinate system.

Each subdomain $\overline{\Omega}^l$ can be easily mapped onto a parallepiped $\hat{\Omega}^l$ by a coordinate transformation (figure (1)). So, cartesian coordinates (x, y, z) in $\overline{\Omega}^l$ can be expressed as functions of curvilinear coordinates (ξ,η,ζ) in the corresponding computational domain $\hat{\Omega}^l$:

$$x = x^l\ (\xi,\ \eta,\ \zeta),\quad y = y^l\ (\xi,\ \eta,\ \zeta),\quad z = z^l\ (\xi,\ \eta,\ \zeta).$$

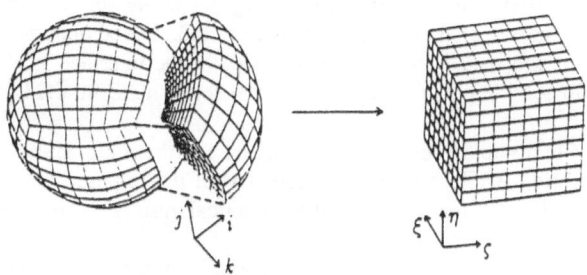

Figure (1) : **Coordinate transformation**

In each subdomain Ω^l, derivatives can be written by using those in the transformed space:

$$\frac{\partial f}{\partial x} = \frac{1}{J}\ \det[\frac{\partial\ (f,\ y,\ z)}{\partial(\ \xi,\ \eta,\ \zeta)}],$$

$$\frac{\partial f}{\partial y} = \frac{1}{J}\ \det[\frac{\partial\ (x,\ f,\ z)}{\partial(\ \xi,\ \eta,\ \zeta)}],$$

$$\frac{\partial f}{\partial z} = \frac{1}{J}\ \det[\frac{\partial\ (x,\ y,\ f)}{\partial(\ \xi,\ \eta,\ \zeta)}],$$

Where J is the Jacobian of the coordinate transformation.

So, we can rewrite equations in the generalized coordinate system in each subdomain.

At interface $\gamma^{m,n}$, derivatives are estimated as linear convex combination of derivatives calculated in its adjacent subdomains.

This matching technique ensures the continuity of all derivatives accross interfaces. It is important to note that no boundary conditions are required on common edges and equations can be 'globally' solved on Ω.

Numerical scheme

The mesh is given in each subdomain Ω^l by a set of coordinates:

$$(x_{i,j,k,l};\ y_{i,j,k,l};\ z_{i,j,k,l}\ )\ |\ i=1,..imax,\ j=1,..jmax,\ k=1,..kmax,\ l=1,..6$$

where *imax*, *jmax* and *kmax* depend on the subdomain Ω^l. We define a new triplet of

133

independent variables ξ, η and ζ such that ξ lines (respectively η and ζ lines) pass through the set of points:

$$(x_{i,j,k,l}; y_{i,j,k,l}; z_{i,j,k,l}) \mid i=1,..imax$$

for each fixed value of j and k (respectively i and k, and i and j) and for each value of l. Surfaces corresponding to $k=1$ and $k=kmax$ are respectively connected to the prolate spheroid surface and the boundary at infinity.

The set of all mesh-nodes in Ω^l is denoted by $\overline{\Omega}_\delta^l$: Ω_δ^l is the set of nodes in the interior of Ω^l and $\partial\Omega_\delta^l$ is the set of nodes on the boundary $\partial\Omega$. The geometrical decomposition is assumed to be conformal. Thus, the set of nodes in the interface $\gamma^{m,n}$ coincides with the set of nodes in $\partial\Omega^m \cap \gamma^{m,n}$ which is also the set of nodes in $\partial\Omega^n \cap \gamma^{m,n}$

Grid points for velocity and pressure are the same. So, there are spurious modes arising from the discretization of ∇q and in conjunction with the imposed boundary conditions. These spurious modes are embedded in the numerical results, so there must be filtered from the physical part of the pressure.

In each subdomain $\overline{\Omega}_\delta^l$ space derivatives are approximated with second order schemes, also on boundaries and interfaces. For a regular function f over $\overline{\Omega}^l$, $f_{i,j,k,l}$ is the value of f at the node $\{i, j, k, l\}$ and we have:

$$(\frac{\partial f}{\partial \xi})_{i,j,k,l} = \frac{f_{i+1,j,k,l} - f_{i-1,j,k,l}}{2\Delta\xi} + O(\Delta\xi^2), \qquad 1 < i < imax,$$

$$(\frac{\partial f}{\partial \xi})_{1,j,k,l} = \frac{-3x_{1,j,k,l} + 4x_{2,j,k,l} - x_{3,j,k,l}}{2\Delta\xi} + O(\Delta\xi^2),$$

$$(\frac{\partial f}{\partial \xi})_{imax,j,k,l} = \frac{3x_{imax,j,k,l} - 4x_{imax-1,j,k,l} + x_{imax-2,j,k,l}}{2\Delta\xi} + O(\Delta\xi^2),$$

where $\Delta\xi$ is the computational grid spacing ($\Delta\xi = 1$).

Quantities $\frac{\partial x}{\partial \xi}, \frac{\partial x}{\partial \eta}, \frac{\partial x}{\partial \zeta}, \frac{\partial y}{\partial \xi}, \frac{\partial y}{\partial \eta}, \frac{\partial y}{\partial \zeta}, \frac{\partial z}{\partial \xi}, \frac{\partial z}{\partial \eta}, \frac{\partial z}{\partial \eta}$ are discretized by the same formulae and then combined together in the usual way to form the metric components.

We denote by ∇_δ^l the approximate derivative operator in domain $\overline{\Omega}_\delta^l$ of the operator ∇ in Ω^l.

• In $\gamma^{m,n}$, derivatives are estimated as the mean value of the two non-centered derivatives calculated at $\partial\overline{\Omega}_\delta^m$ and $\partial\overline{\Omega}_\delta^n$ nodes:

$$\overline{\nabla}_\delta \phi = \frac{\nabla_\delta^m \phi + \nabla_\delta^n \phi}{2}.$$

• In $\partial\gamma_\delta^{m,n}$, derivatives along the interface are estimated as the mean value of the three non-centered derivatives calculated at $\partial\overline{\Omega}_\delta^m$, $\partial\overline{\Omega}_\delta^n$ and $\partial\overline{\Omega}_\delta^g$ nodes:

$$\overline{\nabla}_\delta \phi = \frac{\nabla_\delta^m \phi + \nabla_\delta^n \phi + \nabla_\delta^g \phi}{3}.$$

Thus $\overline{\nabla}_\delta$ keeps the centered second order scheme at interfaces.

Therefore, we denote by ∇_δ the 'global' derivative operator in the set of all collocated nodes in Ω. This operator coincides with the operator ∇_δ^l on subdomain Ω_δ^l $l \in [1,6]$ and with the operator $\overline{\nabla}_\delta$ on all interface nodes.

The time discretization is an Adams-Bashforth scheme for convective terms and an Euler explicit scheme for diffusive terms.
Equations can be written as:

$$
\begin{cases}
A \ (u^{n+1} - u^n) + \delta t \ A_\theta \ \nabla_\delta \ q^{\,n+\frac{1}{2}} = \delta t \ (v \ \nabla_\delta{\cdot}(\nabla_\delta u) - C^*), & \text{in } \overline{\Omega}_\delta \quad (3) \\[2mm]
\nabla_\delta{\cdot}\, u^{n+1} = 0, & \text{in } \overline{\Omega}_\delta \quad (4)
\end{cases}
$$

Where $C^* = (\nabla_\delta \times u^*) \times u^*$ and $u^* = \dfrac{3}{2} u^n - \dfrac{1}{2} u^{n-1}$.

A is a factorized linear operator added to improve the stability properties of the scheme and to correctly damp the numerical non-linear oscillations [5].
A is written in the same form as in the case of the sphere [3]:

$$A = A_\theta \, A_\zeta$$

$$
\text{with} \qquad A_\theta = \begin{cases}
A_{\theta_1} \, A_{\theta_2} & \text{for } l=1,3 \\
\text{or } A_{\theta_2} \, A_{\theta_3} & \text{for } l=5,6 \\
\text{or } A_{\theta_3} \, A_{\theta_1} & \text{for } l=2,4 .
\end{cases}
$$

A_{θ_1} (respectively A_{θ_2}; respectively A_{θ_3}), denotes a periodic tridiagonal solver sweeping subdomains 1, 2, 3, 4 (respectively subdomains 6, 1, 5, 3: respectively subdomains 4, 5, 2, 6) along a particular coordinate. Periodic conditions are imposed with special treatment at the first and last rows of the matrix.

A_ζ denotes a tridiagonal solver taking into account the implicit boundary conditions (figure (2)).

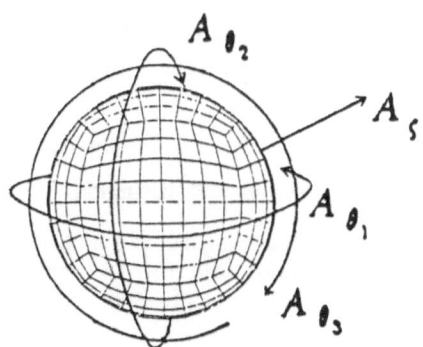

Figure (2) : Domain swept by each operator

These solvers are given by the following formulae:

$$A_\zeta = 1 - \alpha \, \frac{\partial^2}{\partial \zeta^2}$$

$$A_{\theta_i} = 1 - \beta \, \frac{\partial^2}{\partial \lambda_{i,l}^2} \qquad\qquad i=1,..3 \; ; \; l \in [1,6]$$

where $\lambda_{i,l} = \xi$ or η depending on the domain number l.

Second derivatives occuring in A are discretized using a centered second order difference scheme. In order to preserve the accuracy of the time scheme, the parameters α and β are written as [1]:

$$\alpha = \alpha' \, U_\infty^2 \, \delta t^2 + \alpha'' \, v \, \delta t$$

$$\beta = \beta' \, U_\infty^2 \, \delta t^2 + \beta'' \, v \, \delta t$$

where α', α'', β', β'' are constants of the order of unity.

$\nabla_\delta \cdot$ is then applied to equation (3) to give the following equation satisfied by $q^{n+\frac{1}{2}}$:

$$\nabla_\delta \cdot A_\zeta^{-1} \, \nabla_\delta \, q^{n+\frac{1}{2}} = \frac{\nabla_\delta \cdot u^n}{\delta t} - \frac{\nabla_\delta \cdot u^{n+1}}{\delta t} - \nabla_\delta \cdot A^{-1} \, (\, C^* - v \, \nabla_\delta \cdot \nabla_\delta \, u^n \,).$$

At each time step, the expression $A^{-1} \, (\, C^* - v \, \nabla_\delta \cdot (\, \nabla_\delta \, u^n \,))$ is computed by solving the system $A \, \phi = \omega$ where $\omega = C^* - v \, \nabla_\delta \cdot (\, \nabla_\delta \, u^n \,)$. The solution is obtained by solving successive system:

$$A_{\theta_i} \, A_{\theta_j} \, A_\zeta \, \phi = \omega \qquad \text{with: } i, j = \begin{cases} 1,2 & \text{for } l=1,3 \\ or \;\; 2,3 & \text{for } l=5,6 \\ or \;\; 3,1 & \text{for } l=2,4 \, . \end{cases}$$

The LU factorization is done once for all at the beginning of the computation.
Linear system is solved by a multigradient method [6]. Pressure is determined in order to minimize a l^2-norm of $\nabla_\delta \cdot u^{n+1}$.

Results and discussion

The previous Navier-Stokes procedure is applied to a prolate spheroid of axis ratio 1:6 at an angle of attack $\alpha = 30°$ $(u_\infty = (\cos\alpha, \sin\alpha, 0))$. The corresponding Reynolds number, based on the prolate spheroid length $L=2a$, is Re=500.

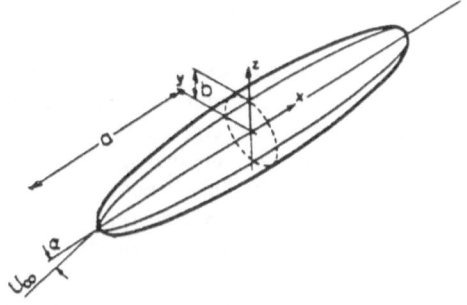

Figure (3) : Notation for prolate spheroid at incidence

The mesh is shown in figure (4). Computational domain is limited by a sphere of radius $r_\infty = 3.5\ L$. The grid contains 45 nodes in ζ direction and 35 or 21 nodes in ξ or η directions and in each sub-domain. Thus the grid system has about 150 000 nodes. Relevant parameters of the mesh are:

$\Delta r_{min} = 5\ 10^{-3}$ in ξ or η direction

$\Delta r_{min} = 10^{-2}$ in ζ direction .

Computation is handled with $\delta t = 2\ 10^{-3}$.

The starting movement is obtained by increasing the free stream velocity till it reaches unity following the law $u_\infty(\underline{x}\ ,n\,\delta t) = (\dfrac{n^2}{n^2 + \lambda})\ u_\infty$ where $\lambda = 100$.

Convergence criterion for the maximal divergence l^2-norm was set to 10^{-4}.

The procedure described in the previous section has been implemented on a CRAY 2 computer. Computational cost is about $7\ 10^{-5}$ second per point and per time step. So, the results required at time $t=5$ correspond to 2500 time steps and the total cost is about 8 hours.

While the flow is symmetric in the leeward-windward plane no symmetric conditions has been used in the computation.

The boundary layer flow may separate on the leeside close to the nose. Separation develops in the form of two symmetric vortex sheets emanating along the separation lines. The pair of vortices run almost parallel to the body upper surface as they grow with distance downstream.

Computed vortices do not appear to lift away from the body surface.

Figure (5) shows velocities in plane $x=0$ and at time $t=5$. For this Reynolds number, no secondary vortex occurs as in the case of higher Reynolds numbers [7,8,9].

The motion seems to be steady and the evolution of the vorticity in time displays the transient state.

Conclusion

A multidomain technique has been described to solve the 3-D Navier-Stokes equations for unsteady incompressible viscous flow. Results show the solver efficiency.

Nevertheless, at the first computational steps, convergence problems were encountered in the pressure resolution. After 100 iterations in the multigradient algorithm, the divergence l^2-norm is only of the order of 10^{-2}. This l^2-norm reaches 10^{-4} for $t > 1$. In fact, the pressure linear system is singular. First, pressure appears as a gradient term in equations so, is defined only up to a constant. Others singularities depend on the finite difference discretisation. To suppress these spurious modes, a solution is to decrease the number of pressure nodes in respect to the number of velocity nodes [10].

There is still work for many improvements such as efficient preconditioners and efficient parallel implementations.

References

[1]   Dang Tran K. and Deschamps V. :"Numerical simulations of transitional channel flow." 5th international conference on numerical methods in laminar and turbulent flow." Montreal, July 1987.

[2]   Mège P. :"Simulation numérique de l'éclatement tourbillonaire par résolution des équations de Navier-Stokes en fluide incompressible." Thèse de Doctorat, Université Paris VI, 1990.

[3]   Le Thanh K.-C., Cantaloube B. and Morchoisne Y. :"Multidomain technique for 3-D incompressible unsteady viscous flow." Numeta 90, Swansea (U.K), January 7-11 1990, Ed. G. N. Pande & J. Middleton.

[4]   Le Thanh K.-C. :"Résolution des équations de Navier-Stokes en incompressible 3-D instationnaire par une méthode de sous-domaines." Thèse de Doctorat, Université Paris VI, 1991.

[5]   Dang Tran K. and Morchoisne Y. :"Numerical methods for direct simulation of turbulent shear flows". Lectures Series 1989-3 on "Turbulent shear flows". Rhode-Saint-Genèse (Belgique).

[6]   Ryan J., Lê T.H. and Morchoisne Y. :"Panel code solvers." Proceedings of the 7th GAMM-Conference on Numerical Methods in Fluid Mechanics, Notes on Numerical Fluid Mechanics, vol. 20, 1988, Michel Deville (Ed.), Vieweg, p. 335-342.

[7]   Meier H. U. & Kreplin H. P. :"Experimental investigation of the boundary layer: transition and separation on a body of revolution." Z. Flugwiss. Weltraumforschung, Vol. 4, p. 65-71, 1980.

[8]   Vollmers H., Kreplin H. and Meier H. U. :"Separation and vertical type flow

around a prolate spheroid, evaluation of relevant parameters." AGARD CP 342, p 14.1-14.14, 1983.

[9] Werlé H. :"Principaux tupes de décollement libre observés sur maquettes ellipsoidales." Note Technique ONERA 1985-7.

[10] Lê T.H. and Morchoisne Y. :"Traitement de la pression en incompressible visqueux." C. R. Acad. Sci. Paris, t. 312, Série II, p. 1071-1076, 1991.

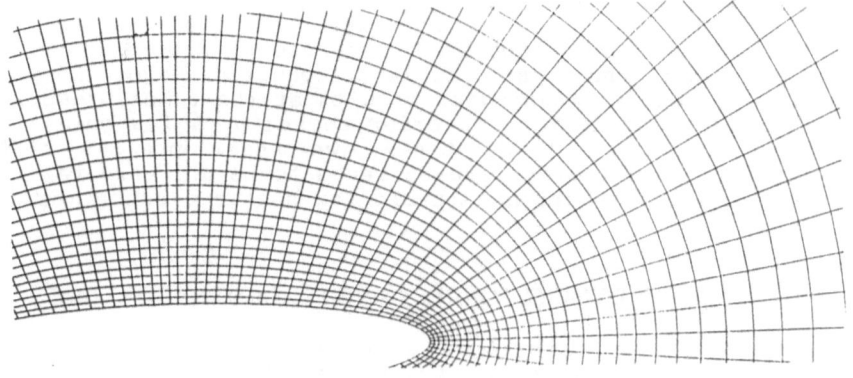

Figure (4): **Partial mesh view at plane z=0**

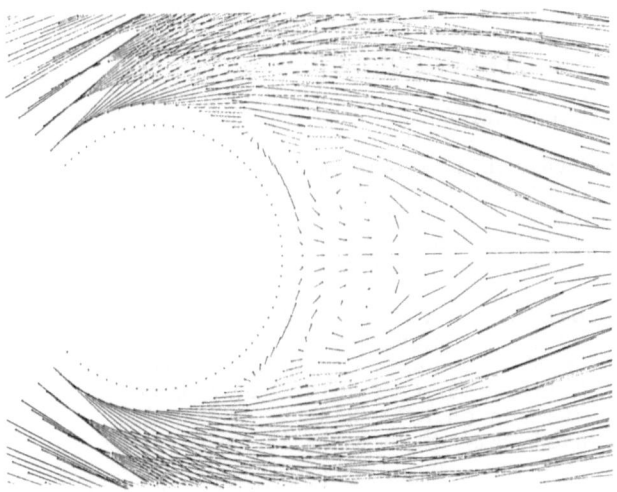

Figure (5): **Velocities at plane x=0 (t=5)**

Final synthesis and concluding remarks

M.O.Deville, T.H.Le, Y.Morchoisne

1 Introduction

This paper summarizes and discusses the main features of the solutions obtained by the various contributors. The comparison will be mostly qualitative because the numerical results do not exhibit the desired quality of converged solutions (in space and time) to be shown as bench mark references. However, we must point out that the test problems required major computer resources to achieve meaningful results at the times prescribed by the problem definition. In order to set up a few guidelines for future work on these tests, we shall present a few quantitative results which are believed to be not very far from a highly accurate solution suitable for use as a bench mark.

2 Three-dimensional Cavity Contributions

We classify the contributions with respect to the numerical method used. They belong to finite differences (FD) or finite volumes (FV) or finite elements (FE). No spectral results were produced even on a regularized version of the boundary condition.

2.1 Finite Differences

[01] Cantaloube and Lê :

A primitive variable approach based on a conservative form is used in a program called PEGASE. The solution scheme is second-order accurate in space and uses centered finite differences. The time integration is performed by an explicit Adams-Bashforth scheme for convective terms and an implicit Crank-Nicolson scheme for viscous terms. The mesh is non-staggered and pressure spurious modes are avoided by eliminating pressure degrees of freedom along the boundaries. An Uzawa-type method is used for minimizing the residual of the continuity constraint. The resulting linear system is solved by a multigradient method. The simulation was run on a 81 x 81 x 231 grid with $\Delta x = \Delta y = 0.0125$ and $\Delta z = 0.013$. The time step Δt is 0.01125 .

[02] Esposito :

The solutions were obtained using a primitive variable approach on a staggered mesh in the spirit of the MAC method ans its numerous variances. The finite differences are second-order accurate, centered except near the walls where they are one-sided. The Adams-Bashforth-Crank-Nicolson scheme is applied for explicit integration of a conservative form of convection and implicit integration of diffusion. A pressure correction algorithm enforces the continuity constraint. The Poisson-like equation is solved by a multigrid method. The computations are performed with a uniform mesh involving 65 x 65 x 193 grid. The time step is fixed to 0.0125 .

[3]Huang, Ghia, Osswald, Ghia :

Here, a conservative form of velocity-vorticity formulation is used on a staggered mesh where velocity components are cell-centered on the faces of the computational grid and vorticity components are defined at cell edge midpoints. The finite difference discretization uses centered schemes both in space and time. The vorticity solver rests upon an ADI scheme while velocity elliptic problems are solved by a multigrid distributive Gauss-Seidel technique. A non-uniform 65 x 65 x 49 grid is employed on one half of the cavity with a symmetry condition applied on the symmetry plane $z = 0$. The mesh sizes are in between $(\Delta x)_{min} = (\Delta y)_{min} = 0.00781$ and $(\Delta x)_{max} = (\Delta y)_{max} = 0.02323$ and $(\Delta z)_{min} = 0.01418$ and $(\Delta z)_{max} = 0.04916$. The time step is equal to 0.005 .

[04] Tromeur-Dervout, Ta Phuoc Loc :

A conservative form of the velocity-vorticity formulation is chosen and discretized on a uniform non-staggered mesh. The space approximation is second-order accurate centered differences while the time scheme is a second-order ADI method. The elliptic velocity problems are solved using a multigrid technique with a plane Gauss-Seidel procedure. A 41 x 41 x 81 grid is used where $\Delta x = \Delta y = 0.025$ and $\Delta z = 0.0375$. The time step is $\Delta t = 0.05$.

2.2 Finite Volumes

[05] Arnal, Lauer, Lilek, Perić :

The velocity-pressure formulation in conservative form is integrated by a finite volume technique with collocated variable arrangement. Centered difference scheme is used for space discretization and two second-order time schemes are employed for time integration, namely, Crank-Nicolson or three time level schemes. The pressure and velocities are coupled through the SIMPLE algorithm. The algebraic equations for the velocity components are solved by the SIP method of Stone while the pressure calculation is speeded up by a multigrid technique. A 32 x 32 x 96 non uniform mesh is used for the full cavity problem while on a 32 x 32 x 48 mesh, the half box problem is treated with a symmetry condition. The average space grid size is 0.03125 with a maximum value of 0.054 and a minimum value of 0.016 . The time step is chosen as 0.05 .

[06] Deng, Piquet, Queutey, Visonneau :

The primitive variable formulation in conservative form is handled by a finite volume technique which uses a cell centered collocated grid. A new interpolation technique, the so-called consistent physical interpolation (CPI), yields a second-order accurate (authors' claim) discrete conservative scheme although a first order upwinding is used for flux reconstruction. The time integration scheme is the second-order three-level Euler backward scheme. Spurious pressure modes are eliminated by the CPI technique. The algebraic system is solved by a fully coupled procedure using a diagonal block preconditioned conjugate gradient method with a CGSTAB algorithm. The test case is computed on a half cavity with symmetry boundary conditions. The grid is uniform and has 65 x 65 x 65 points. The time step is $\Delta t = 0.5$ for $t < 50$ and is reduced to 0.25 for $t > 50$.

[07] Kost, Mitra, Fiebig :

A conservative form of the primitive variable formulation is tackled by a finite volume method expressed on a non-staggered grid. A three time level second order scheme is used as time marching algorithm. The viscous terms are handled by central differencing while convection fluxes are discretized by quadratic upwind differencing scheme (QUDS). Velocity equations are solved by the SIP method of Stone. Pressures are obtained by SIMPLEC algorithm solved by an iterative process. A 25 x 25 x 75 grid is built up with minimum mesh size equal to 0.014 and maximum size of 0.065. The time step is variable ranging from 0.001 for $0 < t < 0.05$ till 1 for $100 < t < 200$.

[08] Mège :

The industrial code PHOENICS is run on the test problem. A velocity-pressure formulation written in conservative form is discretized via the finite volume technique with a staggered mesh. A first-order implicit time scheme is set up. Convection terms are center differenced if the mesh Péclet number is less than two and are treated by a first order upwind scheme if the mesh Péclet number is greater than two. Pressure and velocities are coupled through the SIMPLEST algorithm which is solved by an iterative process. A non-uniform mesh of 50 x 40 x 75 nodes is designed with a minimum mesh size value of 0.005 . The time step is fixed to 1 .

2.3 FINITE ELEMENTS

[09] Janvier, Métivet, M'Gouni, Pot, Razafindrakoto :

The industrial code N3S of EDF is applied to the half 3-D cavity problem. The velocity-pressure formulation is discretized within the Galerkin framework. The time integration uses an explicit method of characteristics for convection. The scheme is first- or second-order accurate. The Stokes solver is based on an implicit first- or second-order Euler backward scheme. The space discretization is performed by tetrahedra with P1-P2 interpolants for pressure and velocities, respectively. The algebraic systems are solved by preconditioned conjugate gradient method. For the pressure, a preconditioned Uzawa algorithm is used. A 41 x 41 x 31 mesh is created with non-uniform nodes distribution. There are 36330 elements with 52661 velocity nodes and 7166 pressure nodes. The time step is 0.02 . Two computations were carried out: the first one up to $t = 200$ is based on first-order time scheme for the advection step; the second one up to $t = 50$ uses the full second-order time scheme.
In order to compare the performance of the N3S code to the finite difference or finite volume approach, we calculate an approximate equivalent "number of nodes" by dividing the total number of degrees of freedom by 4 : (52661 x 3 + 7166) / 4 ≈ 41287 .

2.4 COMPARISON OF THE RESULTS

In order to provide some insight into the computed results, in the next table we will compare:
-the number of Taylor-Görtler-like (TGL) vortex pairs appearing in the transverse direction at the plane $x = 0$ for the times $t = 50$, 100 and 200 ;
-the "performance" on each code, i.e. the CPU time per time step and per grid point (evaluated for the CRAY YMP as the reference computer);
-the CPU time per unit physical time (for t>100., estimated for [03]).

	TGL pairs at t =			"performance" (in s.)	CPU / Δt (in s.)
	50	100	200		
[01]	8	9	9	$2.3\ 10^{-6}$	310
[02]	9	9	9	$8.2\ 10^{-6}$	535
[03]	13	not comp.		$2.0\ 10^{-5}$	830
[04]	7	7	9	$2.5\ 10^{-5}$	70
[05]	0	7	9	$3.0\ 10^{-4}$	590
[06]	9	9	not comp.	$1.6\ 10^{-3}$	1760
[07]	4	7	7	$4.4\ 10^{-4}$	20
[08]	0	8	6	$7.5\ 10^{-4}$	112
[09]	5	7	5	$4.0\ 10^{-4}$	825

2.4.1 RESULT QUALITY

From the experimental results given by Koseff et al. [12,13], we should observe nine pairs of vortices at the reference times. In order to understand why the authors obtain such various results, we have to look at the accuracy problems.

-Time step

The characteristic time scale is varying from the order of unity at the beginning of the integration to 5. for $t > 100$. We can then obtain approximations of the time step for a given value of the accuracy (say 1%). With a first-order time scheme, the time step has to be very low : from .02 at the beginning to .1 for $t > 100$. In [08] the scheme is first order with $\delta t=1$ during all the calculation.

With a second-order time scheme, the time step has to be 5 times smaller at least than the characteristic time scale of the phenomenon :
$\delta t < .2$ at the beginning,
$\delta t < 1.$ for $t > 100$.

We have $\delta t=.5$ in [06] (for t<50.).

-Mesh size

We have to calculate a solution with 9 pairs of TGL vortices plus two corner vortices along the lower x edges (z = ± 1.5 and y = -.5). That means 20 vortices for a length of 3. So if we suppose that 6 points (including "boundaries") are absolutely necessary for an accurate calculation of one vortex, we need roughly speaking at least 100 points in the z direction.
For a constant mesh size (mesh stretching is certainly not necessary), the mesh size in the z direction has to be less than .03 everywhere.
We have $\delta z_{max}=.04916$ in [03], $\delta z=.0375$ in [04], $\delta z_{max}=.054$ in [05], $\delta z_{max}=.065$ in [07], $\delta z_{max}>.05$ in [08] and $\delta z_{max}=.16$ in [09].
Therefore, those calculations suffer from a lack of grid resolution and incur space inaccuracies.

2.4.2 Symmetry

One of the critical issues addressed by various contributors is related to the question : is the flow always symmetric with respect to the plane $z = 0$? ". Several teams assumed from the very beginning that it was the case : [03], [06], [09]. Among the contributors, the answers are quite different. Esposit [02] observes a symmetric solution at every time step, while Cantaloube and Lê [01], Kost et al. [07] point a loss of symmetry during the time integration (in between $t = 50$ and 100 for [07], after $t = 100$ for [01]). Arnal et al. [05] obtain a different symmetric solution if they solve the full problem or the half box with a symmetry condition. They attribute this discrepancy to their implementation of the symmetry condition. Arnal et al. [05] estimate that the space error on the velocity field might be of the order of 10% when they use their 32 x 32 x 96 grid.

2.4.3 Towards a Bench Mark Solution

The influence of time and space discretizations has also been checked by Tromeur-Dervout and Loc after the workshop. They ran their code on the iPSC860 at ONERA. Figure 1 shows the ω_x component in the mid-plan $x = 0$ at time $t = 50$. The mesh is built up on a 40 x 41 x 78 grid and the CFL number is two. One obtains 7 TGL vortices. If we use the same mesh but decrease the time step to a CFL number of 1, we get on Figure 2 at time $t = 50$ eight pairs of vortices . Finally, on figure 3, the ω_x contours are displayed for a run with a mesh of 65 x 75 x 160 grid points and a CFL number of two. We now observe 7 pairs of vortices whose size and strengh are more intense, especially for the middle pair. From those runs, we cannot draw any definite conclusions. The flow is highly transient and advection dominated. Any numerical simulation with standard schemes suffers from numerical errors induced by dispersion and dissipation. This is one of the reason why at the same times, the solution may yield a different number of TGL vortices. To achieve a reasonable accuracy, the computation has to be carried out with fine time steps and fine meshes. These requirements involve the use of efficient solvers on powerful computers.

2.4.4 Method Efficiency

Methods [01], [02] and [04] seem to be well adapted for such simple cases and give us rather accurate solutions.
Methods [08] and [09] are of "industrial" type: they can take into account complex geometries, turbulence models and very general boundary conditions ...etc... So they are more expensive and cannot be used with a sufficiently large number of time steps and grid points.
Between those two groups of methods we can mention method [06] which seems to be accurate but very costly due the complete implicit structure of the time scheme and to the rather complicated space scheme.

3   Prolate Spheroid Contributions

3.1   The Numerical Methods

[10] Deng, Guilmineau, Queutey, Visonneau :

The Navier-Stokes equations, in velocity-pressure formulation, are written in curvilinear body-fitted coordinates. The numerical approximation is designed through an exponential scheme extended to multidimensional problems in the framework of the finite volume technique. The discrete variables are collocated on a non-staggered grid. A pressure equation is obtained through the continuity constraint. This leads to a symmetric, definite positive matrix which is solved by a preconditioned conjugate gradient. The time integration uses a first- or second-order implicit scheme with variable time step. For three-dimensional problems, the momentum equations are solved in a decoupled fashion. The far field condition sets the velocity equal to the free stream velocity. A non-uniform mesh of O-O type is built up with 61 longitudinal points, 50 points in the normal direction to the body surface and 41 points in the circumferential direction. The time step is set to 0.1 . A steady-state solution is obtained around t = 4 .

[11] K.C. Lê Thanh :

A finite difference method is applied to the velocity-pressure formulation of the Navier-Stokes equations. The discrete unknowns are defined on a non-staggered mesh. Second-order centered differences are used for space approximation. An Adams-Bashforth scheme integrates the convective terms while viscous terms are treated by an explicit Euler scheme with an implicit stabilization procedure. The pressure calculation is based on a multigradient technique which drives the residual of the continuity equation to zero.
The surface grid is produced by projection of a rectangular parallelepiped surface grid onto the prolate spheroid. Then, the full mesh is obtained with six subdomains. At interfaces, a matching technique ensures C_1 continuity. The outflow boundary condition is the uniform free-stream velocity. The mesh has 45 nodes in the radial direction and 35 or 21 points in the other directions in each subdomain. The full grid has about 150,000 points. The time step is equal to $2. \ 10^{-3}$.

3.2   Comparison of the Results

Both methods yield qualitatively and quantitatively the same results. The computational cost is increased in this test case per point and per time step because of the complex geometry of the body in the flow. This was of course expected as the presence of coordinate mappings involves more operations per point like Jacobians, non constant coefficients in the derivative operators, etc.
The physics is steady-state and the difficulty is only due to the geometrical complexity and the associated cross-flow separation.

4 Concluding Remarks

Although the participants used major computational resources (number of grid points, small time steps, efficient solvers) on the lid driven cavity, no converged solution has emerged from the workshop. Some numerical results are in good agreement with the experimental data and bring confidence that numerical simulation is indeed a powerful tool. A striking feature of table I is also the discrepancy in terms of computing cost. Some codes because of their generality (geometry, multi-purpose, turbulence modelling, etc) are, roughly speaking, two orders of magnitude more expensive than research tool programmes written to solve the specific problem. It is therefore difficult to make meaningful comparisons with that only criterion in between different codes. While the simple geometry case (and difficult physics) attracted so many research teams, the prolate spheroid problem was only solved by two groups. Is "complicated" geometry so difficult to treat that only a few existing programmes are able to cope with such situations ?

As a final conclusion, we think that much work needs to be performed in years ahead. Open questions are : more accurate schemes, adaptative grids, adaptative time stepping, grid generation, new computer architectures, etc.

References

[01] B.Cantaloube, T.H.Lê,
      "Direct simulation of unsteady flow in a three-dimensional lid-driven cavity".

[02] P.G.Esposito,
      "Numerical simulation of a three-dimensional lid-driven cavity flow".

[03] Y.Huang, U.Ghia, G.A.Osswald, K.N.Ghia,
      "Velocity-vorticity simulation of unsteady 3-D viscous flow within a driven cavity".

[04] D.Tromeur-Dervout, L. Ta Phuoc,
      "Multigrid and ADI techniques to solve unsteady 3D viscous flow in velocity-vorticity formulation".

[05] M.Arnal, O.Lauer, Ž.Lilek, M.Perić,
      "Prediction of three-dimensional unsteady lid-driven cavity flow".

[06] G.B.Deng, J.Piquet, P.Queutey, M.Visonneau,
      "A fully implicit and fully coupled approach for the simulation of three-dimensional unsteady incompressible flows".

[07] A.Kost, N.K.Mitra, M.Fiebig,
      "Numerical simulation of three-dimensional unsteady flow in a cavity".

[08] P.Mège,
     "A 3-D driven cavity flow simulation with PHOENICS code".

[09] L.Janvier, B.Métivet, R.Mgouni, G.Pot, E.Razafindrakoto,
     "A 3-D driven cavity flow simulation with N3S code".

[10] G.B.Deng, E.Guilmineau, P.Queutey, M.Visonneau,
     "Computation of 3-D unsteady laminar viscous flow over a
     prolate spheroid at incidence by a collacated finite
     difference method".

[11] K.C.Lê Thanh,
     "Multidomain technique for 3-D incompressible unsteady
     viscous laminar flow around prolate spheroid".

[12] J.R.Koseff, R.L.Street, P.M.Gresho,C.D.Upson,
     J.A.C.Humphrey,W.M.To,
     "A3-D driven cavity flow: experiment and simulation",
     Proc. 3rd Int.Conf.on Num.Meth. for Lam. and Turb. flow,
     Univ. Washington, pp 564-581, Aug. 8-11, 1983.

[13] J.R.Koseff, R.L.Street,
     "Visualization studies of a shear driven 3-D recirculating
     flow", J. Fluids Engi., 106, 21-29,1984.

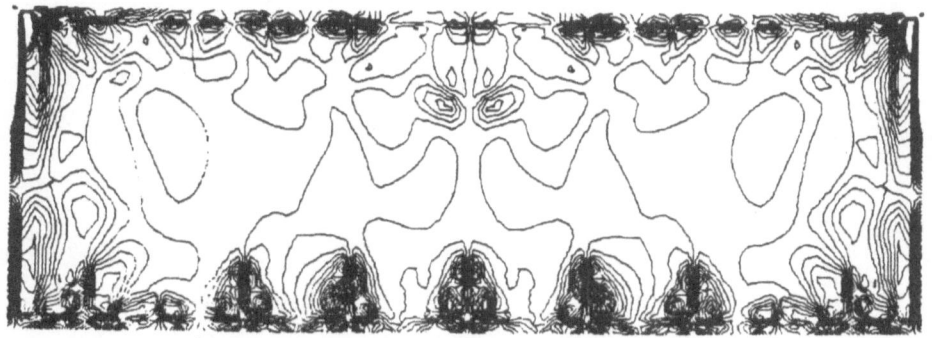

Figure 1 : 40 × 41 × 78 mesh, CFL=2
iso-vorticity, plane x=0 , Jacobi, $\omega_{min} = -5, \omega_{max} = 5$

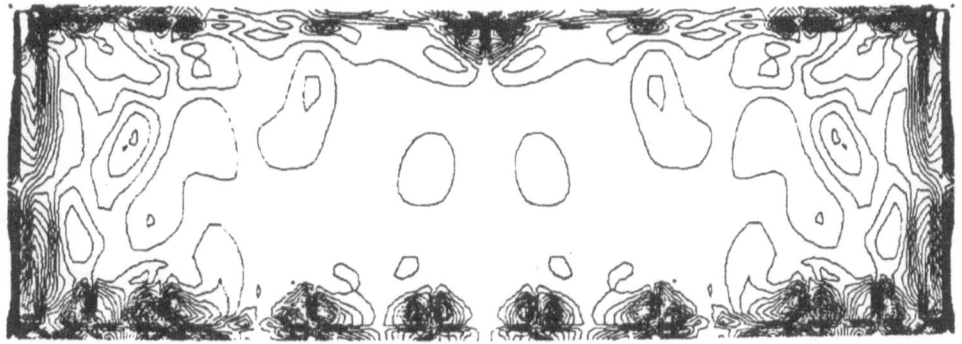

Figure 2 : 40 × 41 × 78 mesh, CFL=1
iso-vorticity, plane x=0 , Jacobi, $\omega_{min} = -5, \omega_{max} = 5$

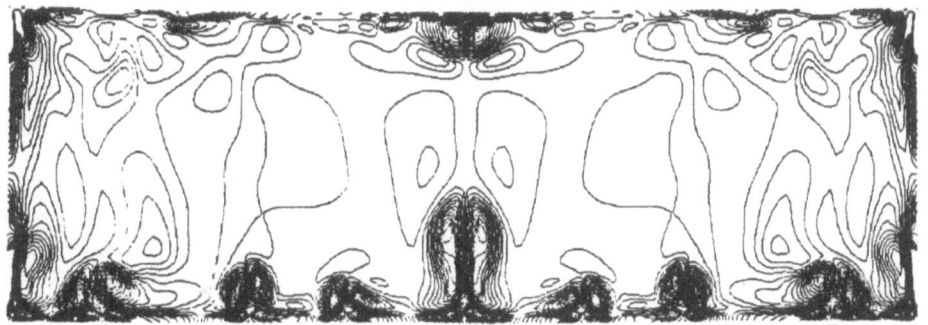

Figure 3 : 78 × 69 × 154 mesh, CFL2
iso-vorticity, plane x=0 , Jacobi, $\omega_{min} = -5, \omega_{max} = 5$

Addresses of the Editors of the Series "Notes on Numerical Fluid Mechanics"

Prof. Dr. Ernst Heinrich Hirschel (General Editor)
Herzog-Heinrich-Weg 6
D-8011 Zorneding
Federal Republic of Germany

Prof. Dr. Kozo Fujii
High-Speed Aerodynamics Div.
The ISAS
Yoshinodai 3-1-1, Sagamihara
Kanagawa 229
Japan

Prof. Dr. Bram van Leer
Department of Aerospace Engineering
The University of Michigan
Ann Arbor, MI 48109-2140
USA

Prof. Dr. Keith William Morton
Oxford University Computing Laboratory
Numerical Analysis Group
8-11 Keble Road
Oxford OX1 3QD
Great Britain

Prof. Dr. Maurizio Pandolfi
Dipartimento di Ingegneria Aeronautica e Spaziale
Politecnico di Torino
Corso Duca Degli Abruzzi, 24
I-10129 Torino
Italy

Prof. Dr. Arthur Rizzi
FFA Stockholm
Box 11021
S-16111 Bromma 11
Sweden

Dr. Bernard Roux
Institut de Mécanique des Fluides
Laboratoire Associé au C. R. N. S. LA 03
1, Rue Honnorat
F-13003 Marseille
France

Brief Instruction for Authors

Manuscripts should have well over 100 pages. As they will be reproduced photomechanically they should be typed with utmost care on special stationary which will be supplied on request.
In print, the size will be reduced linearly to approximately 75 per cent. Figures and diagrams should be lettered accordingly so as to produce letters not smaller than 2 mm in print. The same is valid for handwritten formulae. Manuscripts (in English) or proposals should be sent to the general editor, Prof. Dr. E. H. Hirschel, Herzog-Heinrich-Weg 6, D-8011 Zorneding.